Dedication

In fond memory of and with deep gratitude to the *man's man* we just naturally called The Chief — Dr. Willis Haviland Carrier.

How to Have Air-Conditioning and Still Be Comfortable

HERBERT L. LAUBE

Business News Publishing Company
Birmingham, Michigan

Printed in the United States of America
by
Lithocrafters

Contents

Tables

Figures

Introduction

With a hissing of steam and a clanking of steel the huge locomotive thundered past the station. Behind it clattered the baggage cars, coaches, tourist sleepers and pullmans, their brakes screeching. It was Santa Fe's No. 2—The Navajo—eastbound for Chicago.

Ten hours earlier it had pulled out of Los Angeles. At San Bernardino a second engine had been attached to the rear of the observation car. It had helped to boost the train over Cajon Pass, some 3000 feet above the city. Then downgrade from Summit, where the helper was always dropped, through Victorville, across the Mojave Desert to Barstow. After coasting down the long grade from the desert plateau almost to sea level, The Navajo, with a groan and a sigh, was coming to an impatient halt in Needles. Needles is where the Colorado River cuts between the desert country of California and that of Arizona . . . the hottest point on the hot desert portion of the Santa Fe route. And that is where—as is the case throughout the strip of land from Los Angeles to Phoenix—the hottest weather comes not in July but in September.

After leaving Victorville, the cars had become ever hotter as the relentless September sun climbed higher in the sky and beat ever more strongly on the helpless train and its suffering cargo. As the crack train stopped—on time—it was 7:10 PM. No sooner had the porters opened the doors and placed their footstools than, sweltering, dust-covered and weary, the sweating passengers streamed out. A stoic Indian brave, in beaded leather, was impassively banging away on a huge steel triangle, stridently calling them to dinner.

Between Los Angeles and Kansas City The Navajo carried no diners. It was before the days of the proud Santa Fe Chiefs. Only its fastest train—the all-pullman California Limited—carried diners all the way on its 67-hour rush from Los Angeles to Chicago. The more leisurely Navajo took three and one-quarter hours longer. As did all other Santa Fe passenger trains on that run, it stopped for meals. Passengers welcomed the opportunity to walk about, stretch their legs, get a good meal, quickly served by Fred Harvey's attractive girls.

But not in Needles, not in September. That is where the girls were paid a bonus, because of the heat. That is where the red bricks of the station platform, sunbaked to over 140° during the day, vengefully radiated heat back in infrared form. That is where passengers walking from the rear cars to the Harvey House doors had, at times, actually fainted from heat stroke. And that is where, to keep the girls on the job and counteract adverse publicity, the Santa Fe Railway had first tried something new—air-conditioning. The installation, which served the separate lunch counter and the more expensive, almost sumptuous dining room—linen tablecloths and napkins—had just been completed with my help.

Passengers quickly approached the dining room door where I, joined by the Harvey manager, was standing just inside. As they rushed in each, in turn, suddenly stopped for a double take. They were amazed, astonished, flabbergasted. Unlike the Barstow Harvey House, where they had stopped for lunch, it was cool inside—delightfully cool. After having eaten, they waited until the last "all aboard" before reentering the torrid train. The year was 1925.

As The Navajo disappeared into the darkness, its passengers realized they had all witnessed a miracle. Never during hot weather had they experienced anything like this. Never, not even at the movies or in a department store, let alone in trains, office buildings, hotel rooms nor, not even to be imagined, at home.

Now this miracle is taken for granted. Now what, in 1925, was uncommon luxury has become a common necessity. Now the primary summer function of complete air-conditioning*— cooling—no longer makes news; not when it's doing what it should.

Now, at last, I find myself free of the ties that once bound me to business and profession. Now I can share with others knowledge gained while visiting some 80 different lands, in a modest effort to satisfy the universal quest for comfort. And, finally, I now have the freedom and perspective and above all the time to sit back and reflect on the great but inadequate accomplishments of the comfort industry over the many years I have been a part of it.

As I have been doing this, I have increasingly felt within me the need to speak out on a subject that has caused me increasing embarrassment, both in this country and abroad. Its cause?

---

*As used here the term *air-conditioning* is not limited to cooling. It is the process of maintaining within enclosed spaces those conditions which provide physical comfort for their occupants not just in summer, but throughout the *entire* year.

Very often I am introduced as an authority in the field of environmental control—elder statesman of the air-conditioning industry—superlatives of that sort. Then, shortly afterwards, someone asks, in effect, "Why is it that everyone likes comfort, yet so many people dislike air-conditioning?" This needless and absurd paradox is the cause of my embarrassment or, should I say, *Laube's lament*.

My purpose, then, is to help the reader to draw *comfort* and *air-conditioning* more closely together. The occupant of an air-conditioned space has a right to enjoy his personal idea of comfort whenever he is in it. The buyer of a system is entitled to a design that will give him, and those his system serves, a full measure of comfort at a fair cost.

If this book helps others to reach these worthy goals it will have served its purpose. Additionally, my present role as a constructive dissenter may enhance in a small way the vigor of the industry which has been so good to me, for so long. If so, it will have accomplished even more than I dared hope when I set out to write it.

Herbert L. Laube

The Pines
Camillus, New York

xii

Chapter 1

Do You Control Your Own Comfort?

Surprisingly, I was staying home from the office on a working day. Doctor's orders had me there, mourning the recent loss of an unhappy gall bladder. Toward noon the abrasive nasal twang of a familiar voice came from the kitchen radio. It was Arthur Godfrey, selling soap from New York, on a national hook-up. What he said wasn't registering with me. Wasn't, that is, until—with indignant emphasis—he spat out, "I positively despise air-conditioning."

For a moment this petulant outburst shocked me. Then its irritating cause became clear. It was October, the month for which the National Weather Service of the National Oceanic and Atmospheric Administration—formerly the Weather Bureau—has recorded New York City temperatures as high as 90 degrees. How high it was that day Arthur didn't say. But he was hot—very, very hot.

Why, you say, wasn't his office air-conditioned? That is just the point. It was! But every building manager in New York knows that on September 30th the cooling is turned off for the winter, next morning the heat is turned on. Arthur's office apparently faced south. The low-hanging autumn sun was streaming in. He wanted to open his window. It was sealed tight. The relentless heat was enough to make an even less intrepid soul than Arthur throw discretion to the wind. He had my sympathy.

Was Arthur's attitude unusual? While stopping at the bank a few days ago, I unexpectedly met an old friend. We hadn't seen each other for years. So, we shook hands and exchanged the usual social cliches. Then a question occurred to me. His large company had moved into its own new high-rise office building two years before. "Paul," I asked, "How do you like the air-conditioning in your new offices?" He held his nose, shook his head. It was more eloquent than words.

A few years ago Time, Inc., moved into its plush new office building in Rockefeller Center. Shortly thereafter a memo was issued to all employees. The editors of The New Yorker considered it amusing enough to give it space. Here it is under the New Yorker's heading:

CLEAR DAYS IN THE LUCEMPIRE.

TO: New Building Staffers
FROM: New Building Department
    The controls for the window air conditioning units have understandably confused almost everyone. Here's what you do:
1. Ignore all instructions on valve handle.
2. Flip up small (4″) section of the air conditioning cover plate.
3. Locate black valve handle in recess underneath.
4. Ignore all instructions on valve handle.
5. For WINTER only:
    A. To make room warmer: twist knob counter clockwise (as you look down on it).

    B. To make room cooler: twist knob clockwise.
6. For SUMMER only.
    A. To make room warmer: twist knob clockwise.
    B. To make room cooler: twist knob counter clockwise.
7. Take care not to cut your knuckles on the sharp edges of the cover plate.

The window unit blows air over coils through which water flows. In winter this water is hot; in summer it is cold. Twisting the valve clockwise cuts down on the flow of this water and makes the room cooler in winter and warmer in summer. Hence the apparent contradiction in valve twist directions.

To make this confused situation worse, the definition of summer and winter are pretty arbitrary. As far as the air conditioning is concerned, it becomes summer when the average outside temperature is 50° F. This occurs sometime in March-April, and then the water switches from hot to cold.

Below we have printed the instructions in summary form which you may wish to cut out and fasten with scotch tape to your window unit.

WINTER—Clockwise for colder.
             Counter clockwise for warmer.
SUMMER—Counter clockwise for colder.
             Clockwise for warmer.

As stated, the Time-Life Building is now several years old. As worshipers of progress, can't we expect something better today? Sometimes *yes,* but often *no.* A similar system—presumably with proper markings on the valve handles—was recently selected for New York's gigantic new, twin-towered 110-story World Trade Center. This is the structure featured in the July 1969 Reader's Digest as "the first building of the 21st century." These same towers were characterized by the Editor of Forbes as "Tombstones That Memorialize Monumental Misuse of Mission and Money." It has the same comfort limitations as the Time-Life system. Outrage at this lack of progress compelled me to write a *letter to the editor* about it. These excerpts were taken from the one to the New York Times:

> The World Trade Center will house a heterogeneous group. Women, of course, will want it warmer than men. Britishers will want it cooler than Americans. Mohammedans observe their day of rest on Friday, and usually consider Sunday, when the comfort system is expected to be inoperative, a normal working day. And what about those occupants whose home offices are overseas? If they wish to communicate with them by wire, while their home offices are open, they must be in their own offices at any hour of the night, depending on the overseas location.
>
> Will the comfort system selected by the Port of New York Authority satisfy these exacting requirements? Quite the contrary . . . Part-time comfort, due to inadequacy of the air conditioning system, is the fate today of most building occupants. But systems that provide full-time comfort, by giving each principal occupant a choice of heating or cooling under his own control at all times, are perfectly feasible technically, and without cost penalty. Such systems have for some years been increasingly in successful use in many lesser structures.
>
> It seems a pity that the responsible individuals of the Port of New York Authority were either unaware of such modern systems or indifferent to the comfort of the occupants. "The first building of the 21st century", indeed!

But, you say, I don't work in an office; I couldn't care less. Okay, so maybe you travel. Wayne W. Parrish does. He's the Editor of American Aviation. And, in a recent issue, he had this to say:

The new Century Plaza Hotel in Los Angeles is reputed to have cost $32 Million and it's advertised as "the World's most beautiful hotel". Well, at least it's plenty big and obviously cost a lot of dough. . . .
There was a climate control gadget on the wall but it would take a space engineer to determine how to make it work. I never could. The result was so much gadgetry I never had the right climate. Mostly I was too hot.[1]

More recently another editor devoted several pages in two succeeding issues, to "What's Wrong with Motel Air-Conditioning".[2] He reports typical examples of his own experiences and those of a few personal friends:

An excellent motel . . . But, the heat was pouring from the ceiling fixture and changing the wall thermostat had no effect. The only answer according to the Manager—move to another room.

Too early in the season for the equipment to function properly—we were furnished a small portable electric heater.

Many instances where central combined heating-cooling supplied only heat when cooling was needed or vice versa. No chance for proper control at the unit.

At the next motel it was reported that the entire motel was on "central air-conditioning" but "you have individual thermostatic control". The "but" was the big point, for you could set the thermostat at 80 degrees and still get no heat and the cooling seemed to come on and off irrespective of where the thermostat was set.

That the air-conditioning industry itself is gradually becoming aware of its own deficiencies, is confirmed by the next three quotations, all taken from leading publications of that industry. The first is by Joseph B. Olivieri, a leading Detroit Consulting Engineer:

> . . . Say, guess what. I am on vacation. I am writing this in a motel somewhere in the West. I'll bet you're wondering what kind of nut writes articles while on vacation. Well, you see, I might as well, because I can't sleep! Insomnia? Nope— Air Conditioner!
>
> Tonight for the fourteenth time I have had to make a choice—sleep and sweat; or, awake and cool.[3]

Another editor, under the 42-point title 'WHY?', devoted a full page to the same subject. In it, he asked: "If I can ride in comfort all day in an automobile, why must I be uncomfortable in an expensive motel, which supposedly cost two or three times more per room to build than the automobile?"[4]

And here is still another:

FORGET ABOUT IT!
Or, The Hazards of Room (Temperature) Service

ON TOUR—Midnight arrival at one of a chain of very well known hotel-motels. Outside temperature 45° F. Room temperature on arrival, 55° F.

Wall control had HEAT-OFF-COOL selection, plus thermostat, plus thermometer.

Set on HEAT—75°

Tired, went to bed.

Three o'clock in the morning. Awoke. Cold. Pulled extra blankets onto bed.

Eight o'clock, room temperature now 59° F. Too cold to shower.

Raised thermostat to 85° F. Checked for heat from high sidewall register. Nothing.

Left for day.

Returned about 4 p.m. Outside temperature 70° but very bright sun streaming through curtain wall into room. Room temperature, 78° F., muggy.

Set selector to COOL, thermostat at 72° F. Heard fan kick on.

Changed into swim trunks and went to pool.

Returned to room at 5:45. Opened door. Met by blast of hot air. Room temperature 85°. Checked register; hot air blowing from it.

Rechecked stat and selector: on COOL, set at 72°. Ran stat up and down. Nothing.

Set selector on OFF; air stopped. Set selector on HEAT; no action. Set selector on COOL; hot air again.

Called the front office.

"I'm having trouble making my room comfortable. I can't get any cooling at all and on COOL, I get heating".

"We don't have our air conditioning on."

"Why not?"

"It's only 70° outside. In a few hours it will be 45° and you'll want heat."

"But it's hot in here right now. What about the sun load?"

"The sun load?"

"Forget it."[3]

When it comes to operating mechanical equipment, there is an old saying to the effect that *when all else fails, read the instructions.* This, perhaps, is what some of the unhappy critics failed to do. Here, then, is reproduced the instruction card a friend of mine purloined from his guest room in an Ohio Holiday Inn:

OPERATING INSTRUCTIONS

Heated water is circulated to your room unit from a Central Plant during the heating season. Chilled water is circulated

from a Central Plant during the cooling season—MAY 15th
to OCTOBER 1st.

AIR CONDITIONING IS NOT AVAILABLE FROM
OCTOBER 1st TO MAY 15th

WINTER OPERATION

Set your thermostat to your desired temperature.

Set brown switch to "AUTO" which is automatic; when
the desired temperature is reached, motor will come to a
complete stop and automatically start again when temperature
drops 1-½ degrees.

The toggle switch *must* be in "HOT" position. If the toggle
switch is in "COLD" position, motor will run constantly and
room will become uncomfortably hot.

SLEEPING SUGGESTIONS

At a reasonable time before retiring during COLD
WEATHER turn the brown switch to "OFF" position and
leave the windows closed. When you get your morning call,
turn the brown switch to "MANUAL" AND REMAIN IN
BED FOR ANOTHER TEN MINUTES. Ask the operator
to give you a second call if necessary. When room reaches
comfortable condition turn switch to "AUTO".

SUMMER OPERATION

Call office to make sure we are circulating chilled water.
If answer is YES, set thermostat to desired temperature and
throw toggle switch to "COLD". LEAVE WINDOWS
CLOSED.

On the card my friend wrote: "I never did find the brown
switch".

If you don't work in an office or stop in motels, chances
are you do sometimes eat in a restaurant. Ah, restaurants!
That, too often, is where the lady who leaves the house on
a hot July night without a shoulder wrap makes a chilling dis-

covery. That is where she shares, to a degree, the sensation of a freshly caught mackerel, as he is popped into the deep freeze.

Then, after having dined in goose-pimpled luxury, on what the menu unblushingly identifies as gourmet food, she and her escort depart for the cinema. There, hopefully, one or the other or even both of them may find comfort. Then again, they may not. In most cases one cannot be sure. And so it is, wherever one goes, not just at the movies.

This may be what prompted the editor of a leading architectural journal to write, in part:

> "This bus is air-conditioned,
> Please keep windows closed . . ."

In this emancipated age of ours, it is still somewhat disconcerting to sit in a bus and find the above message engraved on the back of the seat in front. It reminds you rather forcibly who . . . *really* is in charge of your environment.

It is, of course, the bus driver, or the janitor, or the building superintendent. You can't have fresh air because it would throw the system off . . . Thus, though we have *in a general sense* progressed technically towards greater and greater control over our interior climate, in so doing we have on a *personal, individual basis,* given up more and more control to the two or three functionairies named at the top of this paragraph. The reason is that architects and engineers often design mechanical systems to create climates for *areas,* rather than for the *people* who use them.[5]

Well, that last paragraph is partly right. But, as we shall see, there is more to it than that. Most of us spend about 80% of our life within enclosed space—home, office, restaurant, car, etc. Increasingly this space is air-conditioned. What, then, do we mean "air-conditioned"? From the previous examples, you may think it refers to cooling only. But, as used here, it means whatever is needed to keep you comfortable indoors— warmth in winter, coolth in summer and one or the other, as needed, during the changeable weather of spring and fall. Air-conditioning, then, is something everybody should like. Increasingly, though, it is being criticized.

Sometimes this criticism is grimly serious. Then again, it is with a light touch of humor, as when an executive walked into his office in the most fabulous of the fine new office buildings on New York's Park Avenue. In his left hand he carried a small cage in which was a canary. In his other hand was a brick, wrapped in a newspaper. As he set down the canary, he turned to his secretary and said: "When that bird keels over this brick goes through the window".

Why, then, is it that so many people say, "I don't like air-conditioning" when its purpose is to provide complete, personal comfort and everybody, but everybody, likes comfort? And what are the specific causes of the complaints against this industry which, so far, has failed to keep its promise? If air-conditioning is not your dish, chances are it is because of one or more of these reasons:

1. Too Cold or Too Hot—The world has three kinds of people: the *too-hots,* the *too-colds,* and the normal ones, such as you and me. But even we have cause for complaint. In May of 1968, Karel Yaske spoke on behalf of the General Services Administration—landlord of our Government buildings—before a meeting of The Building Research Institute, in Washington. Referring to a survey of their buildings, he confirmed this by saying, "the greatest complaint is temperature."

2. *Cold Drafts*—Coming in from the cold, did you ever hover over a hot radiator? The benign, up-flowing zephyr that bathes your face is never condemned as a draft. But the harsh stream of cold air that drops persistently from the diffuser on the restaurant ceiling, down the bare back of the lady below, is a draft. And it is miserable.

3. *Cold, Clammy Feeling*—"Please turn on the air-conditioning," my wife said. Four of us were playing bridge on a rainy March evening in Fort Lauderdale. Knowing what would happen, I obediently went to the wall thermostat. It's thermometer stood at 75°, as it had all that humid day. I moved the lever to COOL. Five minutes later both ladies begged me to turn it off. Why? The clammy feeling, that we had all felt before the conditioner was turned on, had now simply become a *cold,* clammy feeling.

4. *Wide temperature spread*—The cooling stops. Before it is restarted by the thermostat you are too warm. The heating stops; you are cold before it restarts.

5. *Too Noisy*—Noticeable starting and stopping of fans and compressors is the most objectionable, both indoors and out. Sometimes the quality of the sound is more objectionable than its intensity. The next door neighbor, trying to sleep on a hot July night with his window open, may find the sound of the equipment much more objectionable than does its owner, blissfully asleep in a cool room.

6. *Bewildering Controls*—Wayne Parrish was not alone. Some controls even stump the experts. So if you, too, are sometimes bewildered, do not let this shake your self-confidence.

7. *Can't Have It When You Want It*—"What," I asked the manager of the best address on Chicago's LaSalle Street, "do you do when a valued tenant has to work on a hot July night

or weekend?" "We don't do a damn thing. We would have to bring in an engineer and operate 500 horsepower to cool one office," was his emphatic reply. Most standard office space lease forms list the hours when summer cooling may be had.

8. *Keeps Going*—Can't turn it off when you don't want it.

9. *Sealed Windows*—If you're one of those people who, on a balmy day in May, have the urge to open the window and get a good breath of outdoor air—no matter how foul it is— you'll find me in your corner. But please do not use a brick.

10. *"That Stuffed Up" Feeling*—"The moment I enter an air-conditioned room I choke up," says one. "Every morning, when I get to the office, I sniffle and sneeze," says another. Yes, some people are allergic to sudden temperature changes; others to what comes out of an air-conditioner, irrespective of temperature change. These symptoms are now identified as *air-conditioning syndrome* (page 34).

There are, of course, other reasons for being unhappy with air-conditioning. One such may be an inability to have an inoperative unit or system promptly serviced. These, however, are in a different category from the ten complaints just listed. They are the ones most frequently experienced when the source of the conditioned air is functioning or is in a condition to function normally. These ten complaints will all be covered later in more detail. And so will what, if anything, you can do about them.

Chapter 2

Group Therapy Air-Conditioning

In the beginning, all comfort cooling installations were in the *group therapy* category. They all had two things in common. The places they served were densely populated and, despite the well-known differences in people, the same comfort conditions were, of necessity, provided for all of them. In the case of group therapy installations, there still is no feasible alternative. Gradually, comfort cooling ceased to be a luxurious hot weather novelty and began to be taken for granted. Inevitably, therefore, this form of group therapy led to complaints. People could not agree on what temperature they wanted. In group therapy air-conditioning what, then, is the ideal temperature? It is the one that results in exactly the same number of complaints from people who are too warm—mostly men—as are received from those who are too cold—mostly women.

Lighter clothes for men, sweaters for women, would reduce the number of complaints. But there is more to it than that. The nature of the complaints, many of which are justified, and their causes are not the same for all group therapy applications. It is easy to see why. These span a broad and diverse range: theaters (which came first), churches (last), assembly halls and auditoriums; small and large shops, department stores and shopping malls; restaurants, snack bars, cafeterias and night clubs; railway cars, busses and airplanes—which do offer a measure of individual control; and super markets, which have a problem all their own.

One justifiable reason for complaints is that not all systems are good systems. Some are unbelievably inept. And this is not because the industry lacks the art to do better. After all, the air-conditioning system in the world's most comfortable theater, Radio City Music Hall, New York, was designed and installed almost 40 years ago. Nor are all good systems properly adjusted, maintained and operated. But there is still another factor. In no other country do more people suffer more indifference, more shoddy treatment, more abuse from pampered waiters, clerks and the like than we do in America. And no other people take it, as we do, uncomplainingly. No matter how bad the service or the food, we still leave a tip.

VOLUNTEERS WANTED

Perhaps what we need is more heroes; more men who do complain; more wives who aren't embarrassed when they do it. You, for example, could do your wife, yourself and your fellow man a favor if you complained the next time you find group therapy air-conditioning really bad. I don't mean a nasty crack to the clerk, usher or waiter. I mean taking the time to see the head man—or woman—and then complaining in a fair, firm and friendly manner. But before you do this, you will want to be prepared to discuss the subject intelligently. To help you do so, here is a list of common defects and faults that may cause justifiable complaints. Included are some suggestions for correcting them. Nor are these faults limited to one type of application. They may occur in almost any group therapy job.

1. Warm Spots, Cool Spots—Sometimes you will discover a considerable spread in the temperature of various parts of the same space. This isn't all bad. You can always compromise and sit where your wife likes it best.

Group therapy installations are generally forced air systems. Except for the smallest shops, conditioned air is supplied through air ducts. These ducts have branches leading to side wall *grilles* or ceiling *diffusers,* to maintain uniform comfort conditions throughout the entire area. Each branch has—or should have—a *damper* that is adjusted to supply the quantity of air needed to do this, then locked. This is *balancing.* It is not easy. Closing one damper forces more air out of others; opening it does the opposite. Specialized companies now make a business of balancing air-conditioning systems.

In our country the north side of a building receives no heat from the sun; the southwest corner the most. That is why more cool air must be supplied in summer to the southwest corner. But, if that is done in winter—on a sunny day, particularly—the space will overheat. Thus the same balancing may not be suitable for both seasons. So you might want to ask the manager, "How long since you have had your system balanced?"

2. Hot Spots—These come from local sources of additional heat: coffee urns, hot plates, pizza ovens, popcorn machines, lunch counters, beauty parlors. Best bet is to hood them and add an exhaust fan, as used over your kitchen range. In summer, extra cool air may be blown to such spots; in winter, less warm air. But that would not solve the odor problem. For odor removal, an exhaust fan is needed.

3. Drafts—Side wall grilles are called *registers* when they include a damper for adjusting air volume. They should be at least seven feet—preferably more—above the floor. They should be the *double-deflection* type.

These have long horizontal bars in front, short vertical ones behind them, or vice versa. Both sets are adjustable. Thus, they can vary the direction of the stream of discharged air. They can also concentrate it or diffuse it, thus changing its travel distance or *throw,* or *blow.* They should, and usually can be easily adjusted to prevent drafts. Exception: if, when cooling, the air stream hits a beam or light fixture. Cool air is heavy and wants to drop anyway. Such obstructions give it an excuse for doing so. Note that some improperly adjusted or designed grilles over-blow. The air then strikes the far wall, drops and may cause drafts there.

If air is discharged from an opening, it will draw surrounding air into its stream. We call this *induction.* All wall grilles and ceiling diffusers do this. How much room air can they draw into the stream of conditioned air? That depends on their design and the velocity of the discharged air. Some induce up to three or four times the volume of conditioned air entering the space. Why is this important? Because it affects your comfort. Example:

The cool air from the system is delivered to the room at 55°. Desired room temperature is 75°. Without induction, the conditioned air weighs almost three pounds more per 1000 cubic feet than the room air (Table 6-3). Obviously it wants to drop— right down your lady's back. A *one-to-one induction ratio* means an equal quantity of 75° room air is drawn into and mixes with the 55° stream. Result? The two blend to create 65° air. This 65° air weighs only about one and one-half pounds more per 1000 cubic feet than the room air; thus won't drop as fast. Now let us go to a *three-to-one induction ratio.* Three cubic feet of

75° room air join each cubic foot of 55°conditioned air. What is the temperature of the mixture? Exactly 70°. Now 1000 cubic feet of this mixture weighs less than three-quarters of a pound more than the room air. It has very little incentive to drop. But if it did, it would be far less objectionable than 55° air.

I have had too many cold drafts down my own neck from ceiling diffusers to like them. And usually they smudge the ceiling. This is not for lack of filtering. It is due to smoky room air being induced into the stream coming from the diffuser. But ceiling diffusers are here to stay. Some spaces are so large nothing else will do. Their proper application and adjustment is tricky. As with grilles, they should be backed up with dampers to adjust the volume of air discharged. And they should be adjustable, so as to be able to vary the air pattern. If properly sized and adjusted, they are capable of giving good results, particularly in rooms with 10-foot or higher ceilings.

In winter, cold drafts drop from the windows. If the air-conditioning system is not designed to prevent this, one good solution is low capacity electric heaters. They are designed to be recessed into the window sill or mounted on it.

Then again, some systems supply air through floor grilles. One of our favorite restaurants has one long wall made of glass from floor to ceiling. It faces the swimming pool—nice view and all that. But, perforce, along that wall the cooled air is discharged upwards under the chairs that back up to the glass. There is only one solution to this problem. Sit somewhere else. We do.

4. It's Too Warm, Really—Sometimes everyone agrees that a place that should be cool is obviously too warm. Why? Maybe the system is shut off when it should be on; maybe it is set for heating when it should be on cooling. Maybe dirty filters have cut down the air circulation. Maybe the place installed new, more powerful lighting, delivering twice the heat, without enlarging the air-conditioning system. And maybe the system is overloaded. How many people are in the space? For how many was the cooling system designed?

Some years ago a room cooler was installed in the bedroom of the manager of the Mah Jong Club in Singapore. It was understood that two people would occupy the room. It wasn't long before they reported, "Not enough cooling." The resident Carrier engineer stopped by one night. Holding his nose, he counted twenty-two Chinese, lying head-to-foot and shoulder-to-shoulder on rice mats, on the floor. Indeed there wasn't enough cooling— not the extra 8000 or so Btuh this would have needed; nor enough ventilation air, either.

And here is another reason for not enough cooling: the system may be designed to bring in 100% outside air for cooling when it is cool outdoors. Doing this saves power. If, however, the outside air damper is left in the wide open position on a hot day, it will be impossible to cool the place to the desired temperature.

5. *This Time It's Too Cool*—Chances are the thermostat is set too low. If not, it may be due to *flywheel effect*. Flywheel effect? Yes. By that is meant installing a system purposely too small for the maximum load, to reduce first cost and operating cost. It is done for restaurants, churches, some theaters and other places where the peak load lasts only a short time, up to three or even four hours.

Example: A restaurant caters to a luncheon crowd. They can seat a maximum of 100 people plus 20 waiting for seats. There are ten employees. Is the system designed for 130 people?

No, maybe for 80. Then, around 10:00 AM or so, it is turned on. The thermostat may be set for 72°, 70° or even lower. The air gets cooler and cooler. But not just the air. The tables, the chairs, the walls, the floor, everything in the place gets cooler. The first guests may well be too cool. But by, say, 1:00 PM the room will be up to perhaps 78°. During the lunch hour the room absorbed a lot of heat, warming it from 70° to 78°. It helped the machine carry the load. So, come early if you like it real cool; late if you bring a lady.

6. *Hot Scalp*—High intensity lighting is becoming increasingly popular. Power companies and makers of fixtures and lamps like it. But any baldheaded man who has had to sit for hours under 200 or more foot candles is to be pitied. The air supply grilles for such spaces should be below the light fixtures. The air above the fixtures, being hot, should be exhausted. This reduces but not completely eliminates the problem.

This same radiant effect may bother people in a high domed theater, church or similar structure. There the supply grilles may be ten feet above the floor. Then the cool air tends to stratify below them. The area above the grilles is not cooled; no need to do so. This saves power and reduces equipment size, thus first cost. But, on a sunny day, the air above the grilles becomes hot. It radiates heat on the people below. How to stop this? Install a fan to exhaust the hot air from the upper space.

7. *Cold Feet*—This condition is not uncommon during the heating season. The greater the window area, the more apt it is to happen. Often it is not feasible to supply warm air through floor grilles directly in front of the glass—or cool air either, as already mentioned. Frequent opening of doors to the outside aggravates this problem. Its solution? The grille, through

which air returns from the space to the heating system, should be at or near the floor. Pull the cold strata off the floor and the warm air above it will come down. It has to!

8. *Bad Odors*—If these occur it may be caused by any or all of the following:

a. Not enough ventilation air, a weakness in some theaters and restaurants. Sometimes the outside air is purposely reduced to cut operating cost.

b. No air being exhausted from odor sources—kitchens, toilets, etc.

c. The wall opening or duct through which the outdoor air enters the building is located so that, depending on wind conditions, foul exhaust air from kitchens and elsewhere can enter.

If the system is designed to bring in and condition more ventilating air than is being used, increasing the ventilation should eliminate the first source of bad odors. To eliminate the second, exhaust fans must be added. As for the third, the location of the foul air exhaust or the *fresh* air inlet must be changed.

9. *Noisy*—Once a system is installed, there is not much that can be done to reduce its normal noise level, except to convert a *hard* room into a *soft* one, as more fully explained in Chapter 6. Air passing through grilles and registers sometimes creates noise which can be reduced by readjusting them. Sometimes mechanical equipment is misadjusted or malfunctioning. Fan shafts have end play; belts squeal; pump couplings are misaligned. Here the remedies are obvious.

Normally 50% more air should be circulated during the cooling season than is needed for heating. Use of a two-speed fan motor or variable speed pulleys would permit slowing fans in autumn, restoring normal speed in spring. A small difference in fan speed makes a significant difference in the level of air noises. *Warning:* it may upset the system balance. Even so, doing this is better than it sounds. (No pun.) People are less sensitive to air-conditioning noise in summer than they are in winter.

These, then, are the most repetitive problems that may be experienced with any group therapy application. Should you decide to make a complaint or *offer a suggestion,* perhaps what is said about them here may make you more of an *expert* than the key person to whom you address your complaint.

There are several group therapy applications which have problems unique to themselves. Take, for example, bowling centers. Bowling centers have a heavy cooling load up front. It stops at the foul line. It is wasteful to cool the space between the foul line and the pins. However, cool air may drift to these areas from up front, making it too warm there. If this happens, readjusting grilles or ceiling diffusers may solve the problem. If not, it calls for more careful design of the air supply and return systems.

Other group therapy applications with unique problems of their own are restaurants and super markets.

RESTAURANTS

At 7:45 one October morning last year, I was having breakfast at a U-shaped counter in the best known hotel in downtown Minneapolis. I was chilly. Later I noted that the thermometer on the thermostat was at 68°. This thermostat was next to the opening that led into the kitchen. One waitress shoved the thermostat up to 90°. The next one pushed it down again. A ceiling diffuser over the table in back of me was blowing cold air down my neck.

What was wrong with all this? First, being near the kitchen door, the thermostat could not possibly respond to average room temperature. Secondly, and this is very, very important, the thermostat should have been locked inside of a perforated metal cage, with the key in the hands of the person in charge. That, at least, would have stopped the feuding between waitresses. Finally, the ceiling diffuser needed adjusting.

This brings me to the two reasons why restaurants and related establishments are so often too cool. The first is the flywheel effect, already covered in Item No. 5. The second, and major problem, is the thermostat. Waitresses, bus boys, and the like are working. Kitchens are hot. At work they want it cooler than at rest. Dashing in and out of a hot kitchen, they want it still cooler. Down goes the thermostat. Even if it is under lock and key, the boss may be persuaded to shove it down, to please his help. He just might be better off if he tried harder to please his guests.

But even the best system, operated for best results, can not solve all problems. People dancing quite properly want it cooler than those seated quietly at tables. How do you satisfy both when, on a busy night, dinner tables encroach on the dance floor and, at other times it is entirely covered with banquet tables? So you can not satisfy both the diners and the dancers and can only hope that both will be reasonable about it.

SUPERMARKETS

By far the most frequent complaint in supermarkets is cold aisles and cold drafts, low down, in the aisles. This is usually most severe where there is a concentration of frozen food cases in one area. Why? Because each one acts as its own local cooling system. And it does this, even when cooling is not wanted. A 12-foot single deck frozen food case may remove 3000 Btu from the surrounding space every hour; a four-decker as much as 24,000—two tons of refrigeration. In summer the removal of this amount of heat is enough to cool 600 square feet of floor space.

But this cooling effect is not where you want it—it is too low down.

Of course meat, dairy and produce cases, for unfrozen products, do the same thing but to a lesser degree. Because of this unavoidable loss of heat, the heating system of a supermarket must start operating earlier in the fall and continue to operate later in the spring than would otherwise be necessary. This, of course, reduces the length of the cooling season.

But what can be done about the cold aisles? Proper air movement is the answer. It can not, however, be so much as to disturb the flow pattern of the air in the open display cases themselves. One method, where two tall cases are back-to-back, is to separate them a few inches and raise them above the floor the thickness of a two-by-four. A fan mounted above the two cases draws air from underneath and around them. This air is discharged upwards to mix with the warmer, high level air.[6]

A simpler method is to take advantage of the open trench that usually runs underneath the cases. Of course it is covered in the aisles. This trench carries the wires and pipes that serve the cases. It should be made oversize and connected to an exhaust fan at the rear of the store. Then it serves as an excellent collector and eliminator of the cold air that, otherwise, would cause cold drafts.

Often the heating season compounds this problem of cold

aisles. Roof-top equipment is used to cool and heat the store. To save duct work—and space—the air returning from the store to the equipment is being removed from near the ceiling. The heavy cold air collects on the floor and stays there. There is only one way to correct this condition. Remove the cold air from the floor by means of return air ducts running from the floor to the roof-top equipment.

Next time you experience cold aisles in your favorite supermarket, why not ask the manager what he is doing to correct that condition?

SUGGESTIONS

For all I know you may own or manage an establishment that provides group therapy air-conditioning for its valued patrons. Or you may be planning to do so. Then again, you may only be a harassed taxpayer. And your school district may want to air-condition this hybrid called a *cafetorium*. Or your town plans to build a municipal auditorium to serve as a concert hall, convention center, sports arena and whatnot. If so, these suggestions, pertaining to group therapy systems, may be helpful:

1. Essential Facts—To start with, you need answers to these questions: What is the size of the place? How many people will be in it? What will it be used for? What will they be doing? What percentage will be smoking? Tobacco, that is. Table 5-1 tells how much outside air will be needed, per person. Multiply to get the total. Then consider the lighting. How many watts will it use? What motorized and other heat generating equipment will be operating? Multiply total watts by 3.4 to find out how much heat is given off in Btu's per hour (Btuh). Dryers in beauty parlors give off plenty! Once you have this information you have a basis for checking the proposal, if one has been made. You also have some basis for checking the number of tons of refrigeration required to handle the job by applying these simple rules of thumb:

TYPE OF STORE	FLOOR AREA PER TON
Drug Store with Lunch Counter	120 sq. ft.
Drug Store or Jewelry Shop	160 sq. ft.
Camera Shop or Candy Store	180 sq. ft.
Clothing Store	200 sq. ft.
Furrier's Shop	250 sq. ft.
Ma and Pa Market	300 sq. ft.

2. *Noise Level*—System cost goes up as noise level comes down. During a boxing match in the municipal auditorium noisy equipment is not noticed. During a concert it must be quiet. From the most quiet to the most noisy, applications fall roughly into this sequence: concert halls, opera houses, churches, convention halls, stores, restaurants. In places of assembly, the equipment should be as far from the audience as possible, i.e., under the stage.

3. *Local Codes*—Always check them first. They cover many things. A system that brings in 7.5 cubic feet per minute (cfm) of outdoor air per person may be excellent. But the local code may call for 15. Result: larger equipment, higher first cost, higher operating cost. Probably the code was established without air-conditioning in mind. In that case it may be possible to get a *variance*.

4. *More Comfort Per Dollar*—There are several ways to cut the cost of a system without sacrificing comfort and possibly improving it. For example:

a. The more of the work that is done in the factory and the less on the job, the lower its cost and, usually, the better its quality. This favors factory-assembled *package equipment* over *built-up* systems which are assembled on the job.

b. Systems with air-cooled condensers cost less to operate and usually less to buy than those with water-cooled condensers.

c. Systems that use refrigerated water for cooling cost more to buy and operate than those that let the refrigerant cool directly. But, for some jobs, the latter is not feasible.

d. On large jobs, several package units may do a better job and cost less to buy and run than one large central built-up system. Certainly, in a department store it is well to separate the restaurant area and the beauty parlor from the rest of the store, because the requirements are so different. And occupants of private offices will always appreciate having their own units, under their own control.

e. Don't go off half-cocked by *assuming* electricity for heating will cost you more than gas or oil. First cost is usually less. Operating cost may be. I cannot imagine putting fuel-fired heating in a beauty parlor. So much heat is given off by the driers, etc., that supplementary heat is usually only needed for morning *warmup*.

f. Owners of low, flat buildings (factories, supermarkets, automobile showrooms, etc.) can improve summer comfort, reduce cooling costs and protect built-up roofs by keeping them wet. Specialists now produce spray systems for this purpose. Sometimes these may also be used in place of cooling towers, to serve water-cooled equipment.

5. *Free Cooling*—Many places need cooling in cold weather.

South-facing classrooms, for example, may need it at 35°. When it is 60° or 65° outdoors, or cooler, enough outside air may be brought in to do the entire cooling job, without running the compressor. But it is not *free* cooling. Everything operates except the compressor. Relief dampers to the outdoors, or an exhaust fan, must be provided to get rid of the extra air. Filters must usually be cleaned or replaced more often. The controls must be designed to vary the amount of outside air depending on its temperature. The cooler it gets the less air you need. That makes them complex and costly. And, if the condenser is air-cooled, the power used by the compressor drops with the outside air temperature, while its output of cooling increases. Thus, when cooling is needed least, compressor operating cost is lowest. So don't get stampeded by the proponents of free cooling.

6. *Positive Pressure*—For every 1000 cubic feet of air the system puts into the conditioned space it should bring back only 850 or 900 to the equipment. Why? To maintain a slight *positive pressure* within the conditioned space. This prevents or at least reduces infiltration of unconditioned, unfiltered outside air. And the 100 or 150 cubic feet not returned to the system will have no difficulty leaking out through doors when open, windows and door cracks when not.

7. *Ventilation*—Suggested quantities are given in Table 5-1. Do not depend on open windows or open doors for outside air. Bring it in through the equipment. Then it will be filtered, heated or cooled and dehumidified when it enters the occupied space. And the quantity will be more accurately controlled. Warning: the outside air damper should be motorized. When the equipment is turned off or put in the night or weekend position, it should close. If not, look out for expensive *freeze-ups* in winter: the air-conditioning equipment, if it uses water; your

plumbing if not. And of course, unneeded dampers, when open, waste heat.

As already stated, package units—also called unitary equipment—cost less than a built-up job. But they have other advantages: minimum floor space or ceiling mounting; easiest to install; easy to relocate; single responsibility for results and a factory warranty. Today it means something! And the equipment of most manufacturers is certified by the Air-Conditioning and Refrigeration Institute (ARI) up to capacities of 135,000 Btuh of cooling effect—11.25 tons. This means it must satisfy certain standards and that you can depend on its nameplated cooling capacity and power input. These are not all alike. Take two certified units of 36,000 Btuh (3 tons). Under identical operating conditions, the product of one manufacturer is certified to take 5000 watts, of another 6200 watts, or 24% more, with a correspondingly higher operating cost. ARI publishes a list of manufacturers and their model numbers that comply with its Standard No. 210—*Unitary Air-Conditioning Equipment.*

Chapter 3

Air-Conditioning and Health

In 1775, Sir William Buchan (M.D. Royal College of Medicine, Edinburgh) wrote a six-volume work called *Household Medicine or a Complete Treatise on the Methods of Conserving Health and for Curing and Preventing Illness by Simple Methods.* It was translated into several languages. What follows is retranslated from the second French edition of Dr. Duplanil, Doctor in Ordinary to His Royal Highness the Count d'Artois.

. . . it is a fact that . . . the lack of exercise is not the only reason for damage to health; man suffers most often from the effects of the impure air he breathes. It is not unusual to find an assembly of several people remaining several hours in a comparatively restricted space, probably surrounded by candles which tend to use up the air, and also make it less clean for breathing purposes. The air, which has been breathed over and over again, loses its vitality and becomes impotent to dilate the lungs, resulting in Phthisis and other diseases of the chest so common to sedentary workers. Even the perspiration of a large number of persons assembled together in the same place makes the air unhealthy—the danger becomes even more great if any of them already suffer from a disease of any kind—particularly of the lungs . . .

In cities so many things contribute to the "alteration" of the air that it is not surprising that it often becomes death-dealing to the inhabitants. In enclosed spaces air is not only breathed several times, but is also contaminated by sulphurous fumes, smoke and other foreign matter . . .

Air can become dangerous from several causes. Anything which can alter, to a certain degree, its purity, its heat, its coolness, its humidity, *none of which can be controlled,* makes air unhealthy.

Sir William would be astonished to learn that air-conditioning can and does of course, alter all four of these "none controllable" functions. That is the name of the game. But does doing this, as Sir William suspected, make the air unhealthy?

Comfort is defined as *a state of mental or physical ease, especially one free from pain, want or other afflictions.* A comfortable person is a stress-free person . . . one whose nervous system is not attempting to adjust his body to withstand an unfavorable environment. Perhaps the most simple definition of comfort is to say it is the absence of discomfort. Everybody knows what discomfort is: *1. An absence of comfort or ease; 2. Anything that disturbs the comfort.*

It has been demonstrated that a person in a comfortable environment has more energy than does one who is uncomfortable. It is not that comfort gives you more energy. It is because discomfort robs you of energy. When you shiver with the cold or are afflicted with the heat, it is because your nervous system is trying to tell you something. It is trying to tell you that discomfort is bad for you, but comfort is good.

For years efforts have been made by various researchers to accurately evaluate the health benefits of air-conditioning. How many fewer colds and other respiratory illnesses do workers in an air-conditioned office have, as against those in one that is not air-conditioned? How many fewer absences for reasons of health are there in air-conditioned schools as against those that do not have it? A host of variables other than indoor comfort are involved in studies of this kind. No wonder the results have been inconclusive.

There is evidence that air-conditioning, by filtering pollen and dust from the air, does help many people who suffer from hay fever and asthma. But, in this case, filtering alone, without the other functions of air-conditioning, is also beneficial.

Several independent studies have shown that families living in air-conditioned homes have better dispositions, better husband-wife relations, spend more time together, take fewer vacations and go outside the house for entertainment less often than do families in non-air-conditioned homes. And the wives in air-conditioned homes spend less time in dusting furniture, cleaning walls, woodwork, drapes, curtains and upholstery and have more energy for housework and living. All of which is fine but has no direct bearing on health.

There is, however, one application of air-conditioning for which its health benefits have been conclusively demonstrated. That is in hospitals. There it does more than provide comfort; it has distinct therapeutic benefits. In the case of some illnesses it is the major treatment.

Patients with thyrotoxicosis tolerate hot, humid conditions or heat waves very poorly. Their metabolism is high, and therefore their heat production is excessive. They may be unable to eliminate heat from the body surface as rapidly as it is produced and transported to the skin. They develop hyperthermia or fever, and a tachycardia or rapid heart rate. The demand on the circulation for transport of heat from the interior of the body to the skin surface is increased. The increased body temperature leads to increased cell metabolism, and in turn to still greater heat production. This vicious cycle may threaten life if the cardio-vascular or transport mechanism breaks down. A cool, dry environment favors the loss of heat by radiation and evaporation from the skin and may save the life of a patient.[7]

Richard P. Gaulin, Mechanical Engineer, Division of Hospital and Medical Facilities, Public Health Service, U.S. Department of Health, Education and Welfare has said:

. . . temperature and humidity control would seem to have definite therapeutic value in the treatment of such clinical conditions as—

1. Head injuries or operations which affect the heat regulatory center of the brain.
2. Cardiac conditions which may be adversely affected by the strain of involuntarily increased blood circulation induced in an attempt to dissipate excess heat.
3. Hemorrhagic conditions which reduce the quantity of blood in circulation.
4. Thyrotoxicosis, which increases basal metabolism.
5. Diseases of the liver and kidneys.
6. And, many diseases associated with fever.

In addition to the above, controlled temperatures and humidities and air cleanliness are requisites for the treatment of many allergies.

And writing about the effects of heat on the heart, Dr. George E. Burch, Professor of Medicine, Tulane University, has said:

A hot and humid environment can increase cardiac work as much as strenuous exercise. The environment, therefore, should be made comfortable not only for people with heart disease but also for aged people and for patients with debilitating states or any illness in which thermal regulation should be facilitated. . . . Greater use of air-conditioning in hospitals' cardiac wards will assist toward that end.

Dr. Burch and associates conducted summertime tests in two adjacent wards of Charity Hospital, New Orleans. These tests indicated that patients' hearts did 57% more work in the non-air-conditioned, hot and humid ward than in the air-conditioned ward.

Surgeons who operate in both air-conditioned and non-air-conditioned rooms say the recuperative powers of patients is greater when operated upon in air-conditioned rooms. Additionally, according to a number of authorities, the incidence of postoperative pneumonia is lessened substantially where convalescence is in an air-conditioned environment.

During hot and humid weather, patients in the air-conditioned wards of a partially air-conditioned hospital were quieter and had more rapid physical improvement than those in the non-air-conditioned wards. There seems little doubt that psychological benefits resulting from physical comfort and mental tranquility of the patient are conducive to more rapid recovery.

Doesn't it stand to reason that, if air-conditioning helps people to *recuperate* from illness, it should also be of some benefit in preventing the incidence of illness in the first place?

THE AIR-CONDITIONING SYNDROME

Not everybody is benefited by air-conditioning, even good air-conditioning. At the end of Chapter 1 are listed reasons for disliking air-conditioning. Mostly these have to do with the design of the system or equipment, or the way in which it is or is not operated. These complaints will be discussed in more detail later. But one complaint is largely independent of these factors. It occurs with every kind of application, from the private automobile to the public cinema. I have referred to it as *that stuffed-up feeling*. Increasingly, it is becoming known as the air-conditioning syndrome. A recent health column in the Detroit Free Press by Dr. Peter J. Steincrohn commenced with the statement: *The air-conditioner has complicated the lives of those who consider themselves allergic to cold.* Is there such a thing as cold allergy? Probably, although we really do not know. The source of the symptoms, identified as cold allergy may, in fact, have another source. So far, research has been inconclusive. More is needed. What are these symptoms? Editor Frank J. Versagi tried to find out. Among others, he interviewed Dr. Louis J. Noun, an allergist, of DesMoines, Iowa. And he personally spoke directly with some 30 of his patients. Here, then, are typical responses:

After a round of golf, I have to cool off like a stupid horse before I go into the air-conditioned club room, or I'll stuff up immediately. Sometimes my chest tightens up, too . . . 40 year old woman.

Just before I came here, I walked into an air-conditioned office, and I choked up immediately. I begin to fill up even if I'm exposed to a natural draft or a wind outside on an 80° day . . . 47 year old man.

I can stand air-conditioning if I'm not in it too long. But on a long car trip, say, in an air-conditioned car, I begin to sneeze and wheeze and tear . . . 20 year old girl.

Dr. Noun, as quoted by Versagi, says: "Unless one assumes that all such patients are incompetent to report their own reactions and experiences, it is impossible to ignore the evidence that—for some persons—uncontrolled air-conditioning is a contributing factor in allergic problems." And Versagi expresses the personal view that, "Beyond doubt, some percentage of the population is affected adversely by air-conditioning."

Is there, then, any help for those who wish to be comfortable on a hot day but have an adverse reaction to summer cooling? Dr. Noun recommends, " . . . that the temperature indoors should never be greater than a 10° differential from outdoors. In any event the indoor temperature should never be lower than 78 to 80°."[3]

In the early days of air-conditioning, systems were designed for a maximum *pulldown* of 15°, i.e., 80° indoors with 95° outdoors. Today many systems provide 75° or less indoors when it's 95 or even higher outdoors. Apparently, in the view of Dr. Noun, the symptoms of air-conditioning syndrome are mitigated as the difference in temperature, between indoors and out, is reduced—at least for some people. Another debilitating factor has, however, come to light. This seems to be a fungus that may grow on the cooling coils of an air-conditioner, from where it is carried by the air stream into the occupied space.

As reported in Versagi's publication, Dr. B. J. Szappanoys, M.D., an allergist of Birmingham, Mich., has been doing pioneer work in this field since 1963. He has developed a patented fungicide packaged in an aerosol spray can. It is used to coat the cooling coils of air-conditioning equipment, to kill and prevent the growth of spore-breeding mold. It has already been tried on room coolers, central residential systems, automobile air-conditioners and commercial installations. The same source quotes two Birmingham users of this spray:

> I treat my air-conditioning system and I treat my car (with the spray). As long as I stick to my own system and my own car, I'm all right. But, in any other air-conditioned building, I break out and blister all over my back and arms . . . Ross Mack.

> About three summers ago, we left on a vacation . . . in the first air-conditioned car we ever had. But, every time we turned on the cooling, my 16-year old daughter nearly choked—literally. She couldn't breathe, her chest would get full of whistles and wheezes, her eyes and nose ran. . . . So this summer we took the same kind of a vacation . . . in three or four cars—all but one of which was treated with the spray. That car was air-conditioned also, but my daughter could not ride in it without the same thing happening instantaneously. In the other cars she was fine . . . John Forshew.[3]

A personal communication from Dr. Szappanoys advises that joint research on this subject has also been done by St. Lukes Hospital and Marquette School of Medicine of Milwaukee and the Veteran's Administration Hospital in Wood, Wisconsin. It is reported in the lead article in the Aug. 6, 1970 issue of The New England Journal of Medicine by Drs. E. F. Bonaszak, W. H. Thiede and J. N. Fink.

Is air-conditioning syndrome caused solely by cooling systems? Apparently not. From Livingston, N.J., comes this recent report: "A fungus growth found in stagnant water in the reservoir of a home humidifier has been linked to the illness of a 65-year old patient here, according to doctors at the St. Barnabas Medical Center."[3]

Chapter 4

You Are A Walking Boiler

You use the oxygen in the air to burn your food for heat and power—heat, when needed to keep you warm; power to operate your heart for circulation, your diaphragm for breathing, your body and limbs for locomotion and a wide assortment of work and play activities. If, for prolonged periods, your body receives more heat than it can get rid of, or less than it generates, you die.

Resting nude, your input and output of heat are in balance between 81° and 86°. Clothing lowers this temperature. So does exercise (when you are cold the exercise of shivering helps to warm you). Within narrow limits, your body automatically adjusts for higher or lower temperatures. When it's cold, less warm blood—normally at 98.6°—reaches the surface. This conserves internal heat. When it's hot, you are apt to get red-faced. Why? Because more blood is pumped to the surface, in an effort to get rid of the excess heat generated deep within you. Note that, as more blood is pumped to the surface, your heart must work harder.

Note, too, that it takes more power to pump a certain amount of molasses through a pipe than to pump the same amount of water through it. Why? Because the molasses is more viscous. If you live in the tropics you have thin blood. That makes it easier for your heart to pump, to the surface, the extra blood needed to get rid of your internal heat. If you live in a cold climate, your blood is more viscous, especially in winter. Then, your heart is not called upon to circulate as much as during hot weather.

Now suppose you live in Minneapolis. During the coldest weather you take your golf clubs and hop a jet to Miami or, better yet, San Juan or Acapulco. Immediately, your amazingly responsive control system asks your heart to pump more blood. But your blood is still viscous. Your heart, therefore, is overworked. This helps to explain a phenomenon that has long been observed. People escaping northern winters are more apt to have heart attacks immediately after reaching the tropics than they are after they have been there long enough to give their blood time to become thinner.

YOU ARE A HEAT ENGINE

All heat engines consume more energy than is converted into useful work. Compared with the inefficient engine in your automobile, the diesel engine is highly efficient. Yet, in pulling a train from New York City to Chicago, it consumes three times as much fuel as it would if it could convert the energy in the fuel into mechanical energy at an efficiency of 100%.

You, too, consume more energy than is converted into work, useful or otherwise. As does the engine, you must get rid of the excess, even when it is cold. The engine does this in three ways: by spewing out hot exhaust gas; by means of jacket water which is cooled in its radiator; and by loss of heat from the body of the engine and its exhaust manifold.

You, too, normally get rid of your excess heat in three ways. Heat always flows from a warmer to a cooler body. When your body is warmer than the surrounding air, even though you are clothed, you lose heat to the air. Air exhaled from your lungs is normally at 98.6°. If the air you inhale is below that temperature, you get rid of some excess heat everytime you take a breath. Panting increases that heat loss. But the air exhaled from your lungs is not just warmed, it is humidified by the evaporation of moisture from the delicate cilia and other parts of your lungs, bronchial tubes and nasal passages. Result? Evaporative cooling, as when you blow air over a bowl of hot soup to cool it.

The same evaporative cooling occurs, in widely varying degrees, from the moisture supplied to the surface of your skin by the sweat glands. On a cold day your skin seems dry. But it isn't—not entirely. So, when it is hot, women glow, you perspire and horses sweat. (Race horses, in the tropics, get *drycoat;* they lose the ability to sweat. This kills them. Air-cooled stalls have overcome the problem.) In hot weather, or when working hard in cooler weather, perspiration is by far the most efficient natural means of getting rid of your excess heat.

How much heat you have to get rid of, under various conditions, comes later. For the moment, let us consider something else.

THIS BTU BUSINESS

Get yourself an old-fashioned kitchen match. It will be about 2-½" long, made of white pine with a tip principally of sulfur and white phosphorous. Strike the match. Let it burn until two-thirds of it is a curl of flaming black char. Then blow it out. What did you put into the atmosphere besides some sulfur dioxide and other air pollutants? Heat, you say. How much? One Btu.

Actually, a Btu is an example of the quaint systems of weights and measures we still use in this modern country. The letters stand for *British thermal unit*. But even the traditionalist British are switching to metrics. And, if you have a weight problem, you have heard of calories. Actually, when we talk about food calories we mean kilogram calories. That is 1000 ordinary calories and is written Kcal.

So how much heat is a Btu? It is the quantity of heat it takes to raise the temperature of one pound of water (about one U.S. pint) one degree Fahrenheit.

In the metric system, a kilogram corresponds to our pound. It takes 2.2 pounds to make a kilogram; 1.8°F. to equal one degree centigrade. Multiply 2.2 x 1.8. That equals 3.96—the number of Btu's in one calorie (Kcal). Use four Btu's to the calorie and you will be close enough.

Now let us continue to consider you as an average homo

ACTIVITY	BTU/HR
Sleeping	250
Sitting Quietly	400
Standing Quietly	500
Driving Car in Traffic	600
Light Work, Standing	700
Moderate Work, Standing	900
Walking, with Moderate Lifting or Pushing	1,200
Bowling (actual ball handling)	1,500
Hardest Sustained Work	2,200

Table 4-1—Heat generated by 154 pound Man[7]
(Assumes no rest pauses.)

sapiens, even though no one is ever really average. How much heat does your body give off—*must* give off—every hour? It all depends on what you are doing. The harder you work or exercise the more heat you generate and the more you must get rid of. The American Society of Heating, Refrigeration and Air-Conditioning Engineers (ASHRAE) has done much research to get accurate values. Some rounded off values, for a 154-lb. man, are given in Table 4-1.

If you have a short drive to the office, do sedentary office work, can dictate without pacing the floor and sleep eight hours a night, an average value, for the entire day, of 400 Btu's per hour won't be far off. In 24 hours this equals 9600 Btu's. Divide it by four and you get 2400 calories. If this is the amount of energy supplied daily by your food and drink, you should neither lose nor gain weight.

Some people store up energy in the form of fat even though they watch their calories—sort of. Others, no matter how much they eat, stay thin. These, the Doctors say, have excellent metabolism. To me it seems a bass-ackward concept. Such people, an engineer would say, operate at low efficiency. Much of the energy they shovel in their mouths is eliminated elsewhere without having been used.

YOU ARE A BOILER, TOO

As already mentioned, perspiration—next to a swim or shower —is the most effective means of getting rid of excess body heat on a hot day. But it is not the perspiration that does the cooling. Once, with the temperature above 100, I was working over a drafting board in San Bernardino. To protect the drawing, I placed my handkerchief so it would catch the sweat dripping from my nose. That sweat was wasted. It did no cooling. Only that which evaporated from my body and clothes was helpful. So was the moisture evaporated from my respiratory tract into the air I was breathing. And in both cases the heat of my body was converting liquid water into steam—low temperature steam to be sure, but steam, nevertheless.

How much cooling does it do? Your body gets rid of 1000 Btu's for every pound of sweat you evaporate. (It is called the *latent heat of evaporation.*) And a man working hard on a hot, desert day can produce over a quart (two pounds) of sweat an hour. If he is our 154-lb specimen, his judgment falters and physical deterioration sets in after a loss of four quarts. The loss of eight quarts brings on delirium, convulsions and other adverse symptoms. Another two quarts and his blood becomes so viscous that circulation stagnates and body temperature rises to a fatal level.[7]

The heat your body must lose for you to live is, then, of two types. One is the heat you lose because it flows from your warmer body to the cooler air and objects around you. It is cooling done without the evaporation of body moisture. We call this *sensible* or *dry* heat. When the surroundings are cooler than your body, this fact may easily be confirmed by an ordinary thermometer. Not so the loss of heat resulting from evaporation, as moisture from your body disappears into the surrounding air. This is called *latent,* or *wet* heat. How much of your heat loss is sensible, how much latent, depends on the temperature, humidity and motion of the surrounding air. Table 4-2 gives examples.

From Table 4-2 you observe how an increase in room temperature does *not* increase the total amount of heat your body must get rid of. That is determined by your size and activity. But a room temperature increase of only five degrees makes a significant increase in the amount of heat you must lose by perspiring, breathing and possibly panting. Where people move freely about—walking, bowling—they create their own air mo-

TYPICAL ACTIVITY	*Total Adjusted BTU/HR.	75° ROOM TEMPERATURE		80° ROOM TEMPERATURE	
		Sensible	Latent	Sensible	Latent
Seated in Theater	340	69%	31%	55%	45%
Seated, light office work	400	61%	39%	49%	51%
Seated, active office work	450	56%	44%	45%	55%
Dime Store Clerk	450	56%	44%	45%	55%
Bank Teller	500	50%	50%	40%	60%
Old-Time Dancing	850	36%	64%	29%	71%
Walking 3 mph	1,000	38%	62%	30%	70%
Bowling	1,450	42%	58%	32%	68%

*Adjusted for average mix of average men, women and children; men taken at 100%; women at 85%; children at 75%

Table 4-2—Relationship Between Sensible and Latent Heat Loss of People

tion. This increases sensible cooling. But old-time dancing involved two people in close contact. Even though they may have been moving about, their sensible heat loss is less than that of moving individuals who are isolated, one from the other.

So what does this matter of body heat boil down to? If the energy given off by a sixth grader, squirming at his desk, could

all be converted into electricity, it would keep a 100-watt bulb
lighted. An adult, sitting quietly, would heat a pint of ice water
to the boiling point every 27 minutes or easily keep two 60-watt
bulbs lighted. As for a bowler, while handling the ball, his energy
output would make a blaze of glory out of a large Christmas
tree—he would light up 63 lamps; nine strings of seven bulbs
each!

Chapter 5

The Air You Breathe

The audience was eating dinner at one end of the large hall. Chairs, a platform and rostrum had been arranged at the other end for the talk on air-conditioning which was to follow. It was a meeting of the New York City Section of the American Society of Refrigerating Engineers—one parent of the present ASHRAE.

The men seemed to be less interested in the food than in what was taking place near the rostrum. There a canister-type vacuum cleaner was singing its strident soprano. Instead of sucking, the vacuum cleaner was blowing. A tube, inserted into the discharge end of the machine, led to a flabby mass of thin beige-colored rubber sheeting on the floor. And gradually the sags and wrinkles in the rumpled rubber began to disappear. It was a balloon, filling with air. No, not a child's toy balloon. It was the kind of sounding balloon used by the Weather Bureau. By the time the meeting started the balloon was full, sandwiched between the floor and the nine-foot ceiling above.

Soon after the speaker—dwarfed by the balloon—was introduced, he satisfied the curiosity of the audience, at least partly. The balloon, he explained, now had a diameter of nine feet, seven inches. Thus it contained 462 cubic feet—3472 gallons or 27,520 pints—of air. At normal room temperature, the weight of this air was 34 lbs. Then he turned the tables. What, he asked his audience, did this signify to them? What meaning, if any, did it have? No amount of cajoling brought forth the right answer.

"The air in this balloon," he explained, "is the average amount that each one of you breathes every day. As air-conditioning men you should know this." Then he gave them these additional average values:

Each time you breathe you take in 1.1 pints of air.
Normally you take food and drink only 3 to 5 times a day.
But you take air 17-$\frac{1}{2}$ times a minute, 25,000 times a day.
The weight of your daily food and drink is seven pounds.
Your daily diet of air, at 34 lbs., is almost five times as much.
As long as 35 years ago, it was authoritatively estimated that, in our large cities, your diet of germ-laden dust was 1-$\frac{1}{2}$ lbs.—24 ounces every year.

What, one wonders, is it today?

THE AIR YOU BREATHE

The air you breathe, in its uncontaminated state, is 78% nitrogen, 21% oxygen, 1% argon, not counting from almost nil to over 3% of invisible water vapor (clouds, fog, rain and snow are extra). It also contains minute quantities of carbon dioxide, hydrogen, neon, helium, krypton and xenon. To this we humans, with the help of windstorms, forest fires, volcanoes, the birds and the beasts, have added pollutants galore.

And what is the amount of unweighed noxious gas that comes with it—principally carbon monoxide, oxides of nitrogen, hydrocarbons, sulfur dioxide and photosynthesized smog? In the United States these gasses, by weight, amount to three times as much as the soot, dust and grit you can see and feel—called particulate matter. As one small boy said to his little brother, standing with him in a smog-shrouded slum, "It won't hurt you to breathe, if you don't inhale."

Surely you have heard of the bathing practices of the cleanly Japanese peasant. At the end of his working day the father takes a hot bath—a very hot one. His tub is more of a small, vertical tank than the kind we use. When he has finished, it is mama-san's turn. Then come the children. We flinch as we realize they all use the same bath water. And we can't conceive of drinking our own.

But that is what, in effect, we do with air in a room. We breathe and rebreathe the same air. But not only our own. Also that of everyone else in the same space. And, to it, we add the self-imposed pollution of tobacco smoke. No wonder a major portion of all human illnesses are those of the respiratory tract. Some authorities, counting everything from the sniffles to lung cancer, say it is 90%.

In short, modern man lives at the bottom of a sea of air which is contaminated by mineral, organic and bacterial dirt for which he, himself, is largely responsible. And to live, he inhales it constantly, over 1000 times an hour.

FRESH AIR

In the air-conditioning business we call it OA for *outdoor air*. It may actually be more impure than the air indoors. Nevertheless, it is needed for comfort and health. Our lungs constantly convert the oxygen in the air into carbon dioxide. The amount of oxygen we need to live is, fortunately, small.

Thus, the amount of outside air needed to replace oxygen depletion in an enclosed space is only about two cfm per person. It takes three cfm, however, to keep the carbon dioxide within safe limits. But you wouldn't be happy with that.

Why not? Because the amount of ventilating air that is needed should be enough to dilute body odors to the point where they will not be objectionable. And, if odors from smoking are to be reasonably diluted, it takes about four times more ventilating air than for nonsmokers. In large spaces occupied by only a few persons, such as a private home, enough ventilation is automatically obtained by normal leakage around doors and windows.

In densely occupied spaces or larger buildings, mechanical ventilation is needed. But, when an engineer designs a comfort system for such a building, he calls for the minimum amount of ventilation—the minimum, that is, that he considers to be adequate. Why is he so parsimonious with it when the air is free? Because it isn't free. In winter it runs up the heating bill. During hot, humid summer weather it takes three times more cooling power to cool a cubic foot of outdoor air to the desired temperature than it does to recool a cubic foot of air that is returned from the conditioned space to the cooling equipment.

That is why a system may be designed to handle 75% *return air* and 25% outdoor air. But why bother with the return air? Why not simply cool the outdoor air low enough to *carry the load?* For three reasons: 1. to provide enough air circulation in the conditioned space to prevent the sensation of staleness or stagnation that would otherwise exist; 2. if the air entered the room at, say, 15°—which would be necessary if only 25% as much air were to be used—instead of 60°, it would be so heavy it would cause really objectionable drafts; 3. it would take over 50% more power to operate the refrigeration compressor at the temperature necessary to produce 15° air as against 60° air.

For any given activity, the amount of ventilating air required per person depends on the amount of space available per person and on what is politely called his socio-economic status. Recommendations for average conditions are given in Table 5-1.

APPLICATION	SMOKING	CFM PER PERSON	
		Recommended	Minimum
Apartments	Some	20	10
Barber & Beauty Shops	Much	15	10
Department & Dime Stores	Some	10	7-½
Hospital Rooms	None	20	10
Hotel Rooms	Much	25	20
Meeting Rooms	Heavy	40	25
Offices, Private	Much	30	20
Restaurants	Much	20	15
Schoolrooms	None	7½	5
Theaters	None	7½	5

Table 5-1—Recommended Outdoor Air Requirements[7]

In many states, schoolroom ventilation is subject to state codes. Individual states vary widely on this subject. Their codes call for anywhere from 2.5 to 15 cfm per pupil. The latter is six times the former. They can not both be right.

AIR MOTION

"The Orientals are often credited with inventing man's oldest, most durable cooling device—the fan. As far back as 3000 B.C. fans made of pheasant or peacock feathers mounted on large handles were all the rage in China. But fans . . . still served only to set air in motion—not much help when the air itself is warm."[8]

Well, how much cooling *can* we expect from air motion? One thing is certain—a person wielding a fan generates more heat, thus needs more cooling, than a person at rest. To a considerable degree, the extra heat generated by the user of a hand fan overcomes the benefits of its intermittent breeze. So, in India, they went the Chinese one better. They hung a large screen-like fan or punkah from the ceiling. A servant, called the punkah wallah, kept it in motion. Thus, the master received extra comfort because he generated no extra heat by operating the fan himself.

Next came various forms of electric fans. One such I first encountered in July of 1925, when I spent two nights in the Arizona Hotel in Phoenix. My room had a sixteen-inch portable electric fan. It was a taxi fan. Up to six nickels could be inserted in the meter. For each nickel it was supposed to operate half an hour.

Before retiring, I would place the fan on a chair by the bed, take six nickels from the stack on the night table, insert them in the meter and lie down, stark naked, on the sheet. The cooling breeze promptly put me to sleep. But, the moment the fan stopped two and a half hours later (it cheated in favor of the house) I would awaken. So, investing six more nickels, I would continue my slumber. And so on, through the night. Later, the Arizona was air-conditioned.

ASHRAE has long delved into the three major factors that affect personal comfort—temperature, humidity and air motion.

These studies gave birth to the concept of *effective temperature* (ET). The temperature and humidity aspects of ET are discussed in Chapter 6 and illustrated in Table 6-2. Here we are concerned with air motion. As the motion of the air increases, so does its cooling effect, *provided* its temperature is below body temperature.

Effective Temp. (ET)	Relative Humidity (RH)	AIR VELOCITY, FEET PER MINUTE			
		20	100	300	500
71	20%	80.9° F	81.9° F	83.7° F	85.0° F
71	50%	76.6° F	77.8° F	80.4° F	82.3° F
71	100%	71.0° F	73.0° F	77.0° F	80.0° F

Table 5-2—Twelve Conditions for Equal Comfort[1]

Table 5-2 shows twelve different combinations of temperature, humidity and air motion all of which give a 71 ET. Note that at 20% relative humidity (rh) an increase in air motion from 20 feet per minute (0.23 miles per hour) to 500 (5.7 mph) compensates for an increase in air temperature of 85 minus 80.9 or 4.1°. But, at 100% rh, it compensates for 80 minus 71 or a temperature increase of 9.0°.

A typical Singapore outdoor condition would be 82° and 80% rh for a 78.9 ET, at an air velocity of 20 feet per minute. An increase in air velocity to 500 feet per minute would reduce the ET by 5.9 to 73—a significant improvement. This explains why the trade winds have been such a blessing and the electric fan has, over the years, been so popular in the most humid parts of the tropics.

Incidentally, people in the tropics do not like mechanical cooling unless they can feel the air move; people in the temperate zone, however, abhor cold or even cool drafts. Personally, it depends on where it hits me. My automobile air-conditioning is set to blow gently into my face. It creates the delightful sensation of coasting down a long hill on a bicycle. But a breeze is miserable when it hits me in the back of the neck.

AIR   CLEANING

All air-conditioning systems, from the smallest room cooler to the largest central plant, have air filters. Not generally realized is that the primary purpose of the filter is to protect the equipment itself. Buildup of dirt on cooling coils, fan blades and elsewhere reduces equipment capacity, plugs condensate drains, increases maintenance expense and may cause higher sound levels.

Dirty air filters likewise reduce cooling capacity. They are the major source of service calls. And, when they cause compressor failure—as they can and do—they become an expensive luxury. Yet air filters are easily accessible for inspection, cleaning or replacement on virtually all equipment that does not require a specialized technician for its operation. Where dirty filters cause occupants to complain about the air-conditioning, responsibility is easy to place. It is not the fault of the designer, the manufacturer or installing contractor but of the owner or his building operator.

Claims for the dirt removing efficiency of air filters can be misleading. This applies particularly to the *throwaway* and washable filters used in commercial and residential equipment and room coolers. When the salesman says "this filter will catch 90% of the dirt," it may be an honest statement. But, to be complete, he should add, *by weight.* For normal atmospheric dust this may represent less than 1% of the actual *number* of dust particles. It may consist entirely of particles having a diameter of ten microns or larger (1000 microns equals one millimeter; 25,400 equals one inch). Pollens causing hay fever have a diameter in the range of 20 to 60 microns. Thus, they are apt to be trapped even in filters of low efficiency. Dust causing lung damage is generally under six microns.

The efficiency spread of air filters is immense. It ranges from the type just covered, that removes less than 1% of the dust particles, to so-called *absolute* filters. A major use for these is to catch radioactive dust. They remove, *by number,* 99.98% of all particles down to 0.3 microns, i.e., tobacco smoke.

Filtering efficiency is but one factor to consider in the choice of filters. Dirt holding capacity is important. In the case of cleanable filters, so is life expectancy. And cost, of course, also enters the equation. But of major importance is resistance to air flow, which always increases as dirt accumulates.[9]

Say the token filter in your room cooler is twelve inches by twelve inches. That is one square foot of *face area.* If the unit circulates 250 cfm, you have a face velocity of 250 feet per minute. So you want a better filter. You go all the way to the absolute type. What happens? You won't get 50 cfm out of your unit. The cooling coil will frost up. The compressor will operate under adverse conditions. It may fail, the same as with a dirty filter. Why? Because the resistance to air flow through the high efficiency filter is high. To get 250 cfm through it you would have to increase greatly the pressure generated by the blower in your unit. To do this you would have to increase its speed—if you could—so high it would sound like a police siren—assuming the blower wheel didn't fly apart, which it would. Thus, one powerful reason for using low efficiency filters is not their low cost but the help they give in achieving quiet operation.

So far, my remarks have been limited to two types of filters: those that strain out dirt particles and the impingement type which catches the dirt on a sticky surface as the air follows a sinuous passage.

There is a third type of filter you are now hearing a lot about: *the electronic air cleaner* (EAC). They are normally connected to the standard 115-volt AC power supply. Consumption is on the order of 15 watts per thousand cfm. This is converted to DC of as high as 10,000 to 13,000 volts. As the air passes through an electrostatic field, the dirt particles are given an electrical charge. Then the air passes between closely spaced, often adhesive-coated plates carrying, alternatively, positive and negative charges at perhaps 6000 volts. The dust particles are drawn and cling to these plates. These filters must be shut down regularly for cleaning—usually by means of a stream of hot water. Impingement or strainer type of prefilters are used to catch the bulk of the dirt, by weight—the large particles. After-filters are desirable to catch dirt that may fall from the plates of the electrostatic filter.

A good electrostatic filter has an extremely high efficiency. It can remove smoke particles from the air. It has the further advantage of low resistance to air flow.[9]

The *charged media electronic air cleaner* differs from the one just described. In effect, it combines an ordinary filter, made of glass fibers or plastic, in contact with DC charged plates as above. This creates an electrostatic field throughout the filtering material and draws the dirt particles to it. This type of air cleaner (single stage) costs less and is less efficient than the former, two-stage type.

"How often must I clean or replace it?" is the question most often asked about air filters. The answer is easy: whenever it is dirty! In an East 77th Street apartment house in New York

City they have to be cleaned once a month—by which time they are really foul. In my office, on the outskirts of Auburn, New York, the same filters were scarcely dirty at the end of a year. So it depends on the efficiency of the filter, how much dirt is in the outdoor air and how much is brought into and generated within the conditioned space—carpet lint, etc.

Today our air contamination consists of 25% particulates (matter that can be filtered out) and 75% noxious gasses. None of the air filters described will remove the latter. In my opinion, however, some sulfur dioxide and other gasses are dissolved in the condensed moisture running off of cooling coils. Obviously, this occurs only during the cooling season. Activated charcoal filters—if regularly reactivated or replaced—do have the ability to remove noxious gasses. Their use in comfort air-conditioning systems is, however, extremely limited.

There are two factors concerning electronic air cleaners, about which we do not yet know enough, that deserve to be mentioned. They are ozone and positive ions.

What is ozone? It is a form of oxygen. A molecule of the oxygen you breathe consists of two atoms—chemically O_2. Ozone consists of three atoms, O_3. The high voltage components of EAC's generate ozone. How much? It depends on the make and its condition. Large particles of dirt—or a moth—impinging on the charged plates increase ozone production. But why bring it up? Because, increasingly, instead of being considered beneficial in small concentrations, medical authorities consider ozone harmful to humans. In what concentrations? Apparently as little as one-tenth part per million, for prolonged exposure. How do you know the concentration? It smells somewhat between new-mown hay and garlic. If you can smell it on entering the building from outdoors it is too high! You can not tell if you stay indoors, because it deadens the sense of smell. Eventually, ozone breaks down into ordinary oxygen. It does this more rapidly at high humidities than at low.

The second factor has to do with ions in the air. There are two kinds—positive and negative. Some researchers are convinced that these ions have a significant bearing on the way we feel. They say a preponderance of positive ions has a depressing effect; that it has an adverse effect on the physical and emotional condition of humans. In contrast, a preponderance of negative ions has allegedly an invigorating and stimulating effect on *homo sapiens*. It is believed that EAC's create a preponderance of positive ions.

Years ago a German cigarette factory wanted to supersaturate its air, so that tobacco stems would soften and could then be rolled and used for cigarette filler. To cause supersaturation, i. e. fog, a generator of static electricity had one terminal connected to the fogging nozzle, the other to the ground. Because they carried like charges—either negative or positive—the water particles remained separated.

When the terminals were connected to charge the water particles positively, the workers became irritable and tense, with endless arguments and friction. Reversing the terminals, and thus charging the particles negatively, had exactly the opposite result. A delightful sense of euphoria was then experienced by the workers.

In any case, I am firmly convinced that there are certain illusive characteristics in the atmosphere that do affect us. Further research is necessary to run them down, both qualitatively and quantitatively. I have in mind factors not presently controlled that may account for the difference between outdoor air, at its finest, and the best we can produce indoors. In short, there is still nothing so rare as a day in June!

Chapter 6

The Comfort YOU Want

It's time to make one thing perfectly clear. Air-conditioning was not invented and developed for human comfort. It came into the world wearing work clothes.

Long before it was offered for comfort it was used for an unbelievably broad spectrum of other uses: for making everything from sandpaper to cigarettes, corn flakes to cameras, razor blades to rayon; from coating candy to making capsules; from curing Camembert to controlling chemical reactions; for increasing the output of gold in Africa and milk in Singapore.

Even its name came from industry. In 1906 Stuart W. Cramer, then the leader in the field of Textile Mill Humidification, spoke before the American Cotton Manufacturers Association. He explained that the moisture content of cotton, which had to be high to facilitate spinning, was influenced by the moisture in the surrounding air. By passing that air through humidifying sprays, the cotton would be *air-conditioned*. Thus a new expression was born.

PEOPLE AND PLACES DIFFER

Now you, too, want its benefits. And what you want air-conditioning to do for you depends on who you are, how you are dressed, what you are doing and where you are located.

WHO YOU ARE

Take, for example, General David Sarnoff, founder of RCA. Friends of mine in New York City designed and installed an individual comfort system to serve only his office in Rockefeller Center. General Sarnoff insisted that it maintain a temperature of 65° at all times and that is what is was designed to do. Then, there was the Chairman of another world-famous company, much larger than RCA, also with headquarters in Rockefeller Center. He, too, insisted on a system that would maintain an exceptionally low temperature in his office. To save money, his system was designed also to serve the private office of the President. That luckless individual was almost frozen out of the place. Not until he became Chairman himself, was he able to keep his office at a temperature that gave him comfort.

WHAT YOU WEAR

Years ago in Tampico (also elsewhere) Scottish engineers working there were observed to be wearing heavy tweed suits— vests and all—while going fishing on the Gulf under the tropical sun. After all, that's what one wore when fishing in Scotland. It was the proper thing to do. Obviously, then, what you feel compelled to wear, to establish the image you wish to create, influences what you want your comfort system to do for *you*.

WHAT YOU ARE DOING

When you are working hard, generating 1500 or more Btu per hour, you obviously want a cooler environment than when you are sitting still, watching TV.

WHERE YOU ARE

Obviously, what you want from your comfort system is quite different in Singapore, where it's never been cooler than 66°, than in St. Paul, where it has been 31° below, or in Bogota, where it's never been over 70°, than in Baghdad, where it's been 121°.

I have my own subjective evaluation of the comparative importance of air-conditioning's four major functions, depending on where I am. For residential use, in the four cities mentioned, it is shown in Table 6-1.

All of this is just another way of saying that people are different. They are different as to clothing, activity, diet, metab-

WEATHER CONDITIONS

CITY	Baghdad	Bogota	Singapore	St. Paul
Annual Rainfall, Inches	6	42	95	25
Avg. Min. - Winter	41°	49°	73°	9°
Avg. Max. - Summer	107°	67°	88°	82°
Extreme Min. - Winter	18°	40°	66°	-31°
Extreme Max. - Summer	121°	75°	97°	104°

COMFORT FUNCTION

	Each City Is Given a Total of 100 Points			
Cooling	40	—	55	25
Dehumidification	15	—	45	15
Heating	40	100	—	50
Humidification	5	—	—	1ᴜ

Table 6-1—Relative Importance of
Major Comfort Functions[13]

olism, health and temperament. They are different as to the amount of insulation, in the form of fat, they carry on their bodies. They are different as to age. People over 40 want it warmer than those under 40. And, they are different as to sex. Yet you never heard an air-conditioning engineer say *vive la différence*.

Women generally desire a higher temperature than do men, both in summer and winter. Laboratory tests indicate that, on average, this is only a degree or two. But even this, when women are present, creates problems. Then, too, there is a slight change throughout the day in the temperature at which each individual feels most comfortable. This is because our body is about one degree warmer in the afternoon than in the morning. And this doesn't refer to only the chap who has had a couple of double martinis for lunch. If you have what the British call a chill—we call it a common cold—you may want it warmer than at other times.

As we have seen, you are comfortable when the heat generated within your body is in balance with the heat lost from your body, provided no extra effort is needed by your body to bring this about. No instruments have so far been devised that will tell an outside observer whether you feel comfortable. Comfort is strictly a personal sensation. Of the more than three billion people on earth only one knows how *you* feel. That one is yourself.

TEMPERATURE AND HUMIDITY

For almost 50 years ASHRAE has conducted and sponsored research to determine the effect of temperature and humidity on human comfort. The work is still going on. To me its results prove that there is no such thing as regimented comfort. Each individual must decide for himself what comfort is! Nevertheless, we know that an individual will say he is comfortable under various combinations of temperature and humidity. ASHRAE shows this relationship in terms of *effective temperature* (ET). Table 6-2 shows a limited range of these ET's. They are intended

Effect. Temp. (ET)	RELATIVE HUMIDITY (RH)								
	90%	80%	70%	60%	50%	40%	30%	20%	10%
64	63.5	62.8	62.2	61.6	61.1	60.6	60.1	59.6	59.1
66	65.4	64.7	64.0	63.3	62.7	62.1	61.5	60.9	60.4
68	67.3	66.4	65.6	64.9	64.2	63.6	63.0	62.4	61.6
70	69.2	68.3	67.5	66.7	65.9	65.1	64.3	63.6	62.9
							WINTER		
72	71.1	70.1	69.2	68.3	67.4	66.6	65.8	65.0	64.2
74	73.0	72.0	71.0	70.0	69.1	68.2	67.3	66.4	65.5
76	74.9	73.7	72.6	71.6	70.6	69.6	68.6	67.7	66.8
78	76.8	75.4	74.2	73.1	72.0	71.0	70.0	69.0	68.0
		SUMMER							
80	78.8	77.4	76.1	74.9	73.7	72.6	71.5	70.4	69.3
82	80.7	79.2	77.8	76.5	75.3	74.1	72.9	71.7	70.6
84	82.6	81.0	79.6	78.2	76.8	75.6	74.3	73.0	71.8
86	84.5	82.8	81.2	79.8	78.4	77.0	75.6	74.3	73.1

Table 6-2—Effective Temperature (ET)
(At an air motion of 15-25 feet per minute)[7]

to serve as a comfort index for sedentary individuals, normally clothed.

ASHRAE research indicates that in the temperate parts of the country most such people are comfortable in summer at 71 ET. In winter this drops to 68 ET. In summer, then, you should feel equally comfortable when the thermometer reads 71°, with a relative humidity of 100%; when it reads 76.6° with 50% rh; or, when it reaches 82.7° with 10% rh. This explains why dry air at 80° may actually feel cooler than humid air at 75°. The unshaded portions of Table 6-2 show this relationship for summer and winter.

But there is, of course, more to it than that. For one thing, the most recent research confirms what practical observation has long revealed. A space may feel delightfully comfortable when you first enter it from the hot outdoors. But, as you become acclimated, the feeling of comfort diminishes. You want it cooler. Latest research indicates that, *after three hours of occupancy,* you will feel equally comfortable at 75° and 70% rh as you do at 75° and 30% rh.

Latest research also indicates that this talk of *shock* on entering a cooled room from the hot outdoors is largely nonsense— air-conditioning syndrome excepted. This sudden sensation of coolness is associated with sweat on body and clothes. It takes several minutes for the skin temperature to fall and the moisture to disappear. Even heart patients endure the sudden temperature change with ease. This also applies under the reverse situation— leaving the cool space to reenter a hot one.

IT'S REALLY SIMPLE

The concept of heating is simple. That of cooling is more complex. In the first case, the moisture in the air is safely ignored. In the second case it better not be. Why not? For two reasons: Moisture reduction contributes greatly to summer comfort and it may account for as much as 30% of the entire cooling *load*— or even more. Where does this moisture—this wet heat—come from? Some is given off by people; some is excess humidity brought in with the humid outdoor air, for ventilating purposes; some is humidity from other sources: cooking, bathrooms, drinks, plants, aquariums, and the like. No satisfactory comfort cooling system can be designed without taking this moisture into account.

It was 40 years ago that a bright, self-confident young Carrier engineer took on three of us, on this matter of atmospheric moisture—humidity. It was a damp, dreary, smoggy early winter day in New Brunswick, New Jersey. The three of us had our desks in a steam heated open office large enough for ten or twelve people. Why the Carrier man was there I do not recall. But my company—Brunswick-Kroeschell—had just been merged with his—Carrier Engineering Corporation.

In any case, our steam heated office was extremely hot. The air seemed dry. So one of us opened a window. By letting in the moist, almost foggy air, we expected not only to cool but to humidify our space—bringing up the rh to 30 or even 40%. The Carrier man said, "No, your air will simply get drier." We laughed—it sounded ridiculous to us—and we opened another window.

But the Carrier man insisted we had been wrong to open the windows. And he pulled from his briefcase a 15″ x 11″ plastic prop—the Carrier Psychrometric Chart—to prove his point. With a pencil he traced the applicable curves. Outside of one window was a thermometer. It showed 25°. Obviously, the humidity of the outdoor air could not exceed 100%. We agreed to use that figure.

The Carrier chart showed that one pound of air at 25° and 100% rh contained just under 20 grains of moisture. We also agreed that in the office the outdoor air would be heated to 75° —the office was then at least 80°. What would the relative humidity of that 25° outdoor air be at 75° indoors? The chart showed only 17%. We lost the argument. Had the rh at 80° been only 20%, simply turning off the radiators until the temperature dropped to 75° would automatically have increased the rh to 30%.

HERE'S HOW YOU DO IT

Now, let us move from the B-K office of 40 years ago to a small bedroom in your home. Its size? Ten feet by thirteen-feet, eight-inches with a seven-foot, four-inch ceiling height. Why? Because it then has a volume of 1000 cubic feet—an easy quantity to deal with. (And if the adjoining living room is 13 feet, 8 inches, by 20 feet, it will contain 2000 cubic feet.) Now let's chuck grains of moisture—of which there are 7000 per pound of air—and talk about pounds or pints of water in the bedroom air.

At sea level, one pound of air at normal room temperature fills about 13.5 cubic feet. Thus, at around 75°, the air in the empty bedroom weighs 1000 divided by 13.5 or 74.3 pounds. (In mile high Denver, it would be 62 pounds—one-sixth less.) How much water is in that air? Table 6-3, opposite 75°, gives you the answer in pounds or pints for various relative humidities. In summer, with 75° indoors, you will be comfortable at a relative humidity of 50%. Under these conditions, how much water is in the bedroom air? Follow Column 8 in Table 6-3 down to the line opposite 75°. You will see it is 0.7 lbs. That is 11.2 ounces, enough to fill an oversized highball glass. Add the living room and you have three such glasses full.

In a house of average good construction, with windows closed, you have a normal in-and-out leakage of air around the doors and windows. (It is called infiltration on the windward side of

1	2	3	4	5	6	7	8	9	10	11
Air Temp. °F	lbs. per 1000 cu. ft.	RELATIVE HUMIDITY OF AIR IN % RH								
		100	90	80	70	60	50	40	30 *	10
0	86.4	.070	.063	.056	.049	.042	.035	.028	.021	.007
5	85.5	.090	.081	.072	.063	.054	.045	.036	.027	.009
10	84.6	0.11	.099	.088	.077	.066	.055	.044	.033	.011
15	83.7	0.14	.126	.112	.098	.084	.070	.056	.042	.014
20	82.8	0.18	.162	.144	.126	.108	.090	.072	.054	.018
25	82.0	0.22	.198	.176	.154	.132	.110	.088	.066	.022
30	81.2	0.28	0.25	.224	.196	.168	.140	.112	.084	.028
35	80.4	0.34	0.31	0.27	0.24	.204	.170	.136	.102	.034
40	79.6	0.41	0.37	0.33	0.29	0.25	.205	.164	.123	.041
45	78.8	0.50	0.45	0.40	0.35	0.30	0.25	.200	.150	.050
50	78.0	0.60	0.54	0.48	0.42	0.36	0.30	0.24	.180	.060
55	77.2	0.72	0.65	0.58	0.50	0.43	0.36	0.29	0.21	.072
60	76.4	0.85	0.77	0.68	0.59	0.51	0.43	0.34	0.25	.085
65	75.7	1.00	0.90	0.80	0.70	0.60	0.50	0.40	0.30	.100
70	75.0	1.19	1.07	0.95	0.83	0.71	0.60	0.48	0.36	.119
75	74.3	1.40	1.26	1.12	0.98	0.84	0.70	0.56	0.42	0.14
80	73.7	1.65	1.49	1.32	1.16	1.00	0.83	0.66	0.50	0.17
85	73.0	1.93	1.74	1.54	1.35	1.16	0.97	0.77	0.58	0.19
90	72.3	2.25	2.02	1.80	1.57	1.35	1.13	0.90	0.68	0.23
95	71.6	2.63	2.37	2.12	1.84	1.58	1.32	1.05	0.79	0.26
100	71.0	3.07	2.76	2.46	2.15	1.84	1.54	1.23	0.92	0.31

*For 20% rh, multiply 10% rh by two.

Table 6-3—Pounds of Water in 1000 Cubic Feet of Air[7]

the building, where it comes in, exfiltration on the leeward side, where it goes out.) How much? About one air change an hour. So once every hour 1000 cubic feet of outside air enters the bedroom. Let us assume it is a hot day—95° outdoors but with the same relative humidity as you want indoors—50%. How much water is in 1000 cubic feet of that outdoor air? Follow Column

8 down to 95°. It is 1.32 pounds. But you only want 0.7 pounds. So your air-conditioning system must remove 1.32 minus 0.7 or 0.62 pounds. That is about 10 ounces. Still a good glass full. And you must remove it every hour.

But that is not all. If you are in the room you, too, are adding moisture. How much? If you are sitting quietly, two ounces per hour. Where did I get that? Table 4-1 shows 400 Btu per hour as the heat you give off. Table 4-2 shows that, at 75°, 31% is wet heat. That is 124 Btu per hour. You lose 1000 Btu when your body evaporates one pound of water—16 ounces. Since 124 is 12.4% of 1000 and 12.4% of 16 ounces is 1.98 ounces, that is what you lose. I rounded it off to two ounces!

So, to maintain the room at 50% rh, your air-conditioning must remove ten plus two or twelve ounces of water *every* hour. And, if there are other indoor sources of moisture, it, too, must be removed.

Now let us switch to winter. It is a cold day, 10° and snowing. This implies an outdoor relative humidity of 100%. Again you use Table 6-3. Go down Column 3 to 10°. The outdoor air contains 0.11 pounds of water in 1000 cubic feet. Let us say you want an indoor relative humidity of 30% in winter. Opposite 75°, Column 10 shows 0.42 pounds. The outdoor air has 0.11 pounds. So you must add 0.42 minus 0.11 or 0.31 pounds— say five ounces every hour to the bedroom air. That is right, isn't it? No, it isn't. Why not? Because you are still giving off your two ounces every hour. So your humidifier needs only to add three ounces. And, if you have other indoor sources of moisture, it's even less.

The control of moisture in the air—psychrometry—is the hard part for most people to understand. The easy part, which is handled in much the same way, is ordinary, i.e., dry or sensible heat. It flows into your room in summer and must be removed for coolth. In winter it flows out of the room to the cold outdoors. Thus, for warmth, heat must be added. The quantity must, in each case, be calculated for assumed maximum outdoor and indoor conditions. Table 6-4 gives suggested *design conditions* for a number of cities.

This is one point it will be helpful to keep in mind: all heat—whether wet or dry—that is generated indoors *helps* the heating system by doing part of its job; *hinders* the cooling system by adding to its job.

DEHUMIDIFICATION

Late in the fall of 1902 Willis Carrier was waiting for a train in Pittsburgh.

It was evening, the temperature was in the low 30's, and the railway platform was wrapped in a dense fog. As Carrier paced back and forth, waiting for his train, he began thinking of the fog. As he thought he got 'the flash of genius', as patent experts put it, that eventually resulted in *dew-point control,* which became the fundamental basis of the entire air-conditioning industry.[10]

Shortly before his death in 1950, Carrier dictated a memo which went along these lines:

Here is air approximately 100% saturated with moisture. The temperature is low, so, even though saturated, there is not much actual moisture. There could not be at so low a temperature. Now, if I can saturate air and control its temperature at saturation, I can get air with any amount of moisture I want in it. I can do it, too, by drawing the air through a fine spray of water to create actual fog. By controlling the water temperature I can control the temperature at saturation. When very moist air is desired, I'll heat the water. When very

CITY and Location	WINTER			SUMMER		
	DESIGN TEMP. °F	WIND VELOCITY mph	Heating DEGREE DAYS	DESIGN TEMP. °F	RELATIVE HUMIDITY %	DAILY RANGE °F
1*	2*	3*	4*	5*	6*	7*
Atlanta	17	15	2961	95	47	19
Birmingham	18	6	2551	97	45	21
Boston	5	15	5634	91	51	16
Buffalo	0	10	7062	88	55	21
Chicago (O'H)	−5	10	6639	93	49	20
Cincinnati	7	6	4410	94	50	21
Cleveland	2	10	6351	91	51	22
Dallas	17	15	2363	101	38	20
Denver	−4	6	6283	92	22	28
Detroit (M)	3	10	6293	92	49	20
Houston	27	10	1396	96	50	18
Jacksonville	28	6	1239	96	50	19
Kansas City	2	10	4711	100	40	20
Los Angeles	41	3	2061	94	38	20
Memphis	16	6	3232	98	46	21
Miami	43	6	214	92	60	15
Minneapolis	−13	6	8382	92	53	22
Montreal	−14	10	8203	88	58	18
New Orleans	30	10	1385	93	60	16
New York (LG)	10	15	5219	93	49	16

1*	2*	3*	4*	5*	6*	7*
Philadelphia	9	10	5144	93	52	21
Phoenix	31	3	1765	108	25	27
Pittsburgh	5	10	5500	90	50	19
San Antonio	24	6	1546	99	37	19
San Francisco	35	6	3015	83	38	20
St. Louis	2	10	4900	98	44	21
Seattle (B)	23	6	5145	82	46	23
Toronto	-5	10	6827	90	56	22
Washington	14	10	4224	94	53	18
Winnipeg	-30	10	10679	90	50	23

1* Values are for airport locations except Cincinnati, Los Angeles and Pittsburgh. Downtown temperatures usually run 2 to 3° higher.

2* While extreme temperatures are lower (1% of the time), design temperatures are a fair compromise between comfort and system cost.

3* Shown for comparative purposes only. For the same design temperature, more heating capacity is needed with high wind velocity than with low.

4* Indicates, comparatively, the amount of heat required during the year when outdoor temperatures are below 65° F. Thus, if the mean outdoor temperature for one day is 65° it has zero degree days; if 55° it has 10 degree days, if 45° it has 20 degree days, etc.

5* Values are a fair compromise between comfort and cost and are exceeded only 1% of the time.

6* These relative humidities apply only to design temperatures shown.

7* A rough indication of the usual difference between day and night temperatures, during warmest month.

Table 6-4—Comfort System Design Data[7]

dry air is desired, that is, air with a small amount of moisture, I'll use cold water to get low temperature saturation.[10]

Two years later Carrier was issued U. S. Patent #808,897 which, among other things, covered the use of a water spray or mist to dry air. *Dry* air with water? The idea was greeted with incredulity and even ridicule. But it worked. Let us see why.

But first, what is dew point? It is a short way of saying *fully saturated with moisture, or 100% relative humidity.* A dew point of 60 means the air contains so much water that at 60° its rh will be 100%. In short, at 60° it will be holding all the water it can. Spray it with 60° water and still it can hold no more. Spray it with colder water and the air temperature will drop and moisture will be condensed out of the air—dehumidification. Spray it with warmer water and the air temperature will increase and moisture will be added—humidification.

On a hot July day, make yourself a glass of iced tea or a highball. Watch it closely. Soon a faint dew of moisture appears on the outside. No, the glass is not leaking. The moisture is coming from the air. Keep on watching the glass. The tiny beads of moisture grow larger. The moisture itself is taking more moisture from the air. And, if you didn't put a coaster under the glass, you will soon be in trouble with your wife.

Now mix yourself a second drink. Start with a dry glass. Put a thermometer in it. Keep stirring to assure uniform temperature. Watch closely until the first fogging appears on the outside of the glass. Quickly read the thermometer. That is the dew point of the air in the room.

Let us say the fog appeared at 55°. Then, from table 6-3, opposite 55° in Column 3, 100% relative humidity, you will note 1000 cubic feet of air contains 0.72 pounds of water. But your room is at 75°. So, go to the right from 75° in Column 1 until you get 0.72. It doesn't show. But 0.70 does show—for 50% rh in Column 8. Thus, the relative humidity in your room is just a smidgeon over 50%.

If you gave 75° and 50% rh to an air-conditioning engineer he would look at his *psychrometric chart* and say your *wet bulb temperature* is 63°. To verify this, pull a tiny silk sock snuggly over the bulb of an ordinary thermometer and dip the sock in water. Then, with a suitable handle, twirl the thermometer for two minutes, as a child twirls a 4th of July sparkler. Read it quickly. If, in a 75° room, it still shows 75°, the rh is 100%. If it reads 70°, the rh is 80%; if 65°, about 60%; if 60°, about 40%, and so on. This device for measuring wet bulb— as against dry bulb—temperature, and thus rh, is called a *sling psychrometer*.

Early comfort systems used water sprays—i.e., *air washers* —not only to humidify, but to cool and dehumidify the air. It was all done by regulating the temperature of the water—dew point control. And this method is still used for industrial applications requiring accurate humidity control; no longer, however, for comfort systems. There the use of *fin coils* is the rule.

Typically, a fin coil would use half-inch straight copper tubes connected to form a continuous channel through which refrigerant can flow. The entire coil might be *three rows deep*. Crosswise of the tubes would be the fins, made of thin aluminum sheeting. A coil 2.5 feet long might have 360 fins which give the coil much greater cooling power than if made with bare tubes only.

The coil removes both dry (sensible) heat to cool the air forced through it and wet (latent) heat to dehumidify or dry it.

How much of each it removes depends on the design of the coil, the temperature of the refrigerant (or chilled water) passing through its tubes and the temperature and humidity of the air being treated. The advantages of the fin coil (or *evaporator,* because in it liquid refrigerant evaporates into gas) over the air washer is that it takes less power and space and costs less. And it is quiet.

What about the comfort capabilities of the fin coil? It, too, has a dew point temperature. The dew point temperature of a fin coil depends on the temperature of the refrigerant, the design of the coil and the temperature, humidity and velocity of the air passing through it. Under usual operating conditions, its dew point is in the normal comfort range for summer cooling. But, under extreme conditions, the fin coil may remove less moisture than is desirable for optimum summer comfort. There are means —*at slight extra cost*—for correcting this weakness.

So far I have assumed that dehumidification would be accomplished by use of a refrigerating machine. There are two other ways in which it can be done, called absorption and adsorption. Their use for human comfort is too limited, however, to justify discussing them here.

WINTERTIME HUMIDIFICATION

Someone has said the difference between humidifying the air in winter and not doing it is the difference between blowing your nose and picking it. Again, there is more to it than that. At least there was in the speech I made in Albany, New York, one evening in the fall of 1932. Over 100 plumbing and heating contractors formed the audience. My job was to get them to sell *The New Carrier Home Humidifier.*

To hear me tell it, life without a humidifier was scarcely worth living. Up to 90% of all human illnesses involve the respiratory system. Most of it occurs in winter. Why? Without a license to practice medicine, I told them. And the furniture! It was simply a shame, the way the drying out of the wood in winter would make it fall apart. And the floors would creak. Carpets and drapes would be damaged. Lack of moisture would make their fibers brittle. Thus, deterioration would accelerate. Then, too, after walking over such a carpet, not only you but your helpless women and children would get, due to static electricity, a sudden, painful shock on touching each other or a doorknob. After scaring them all I could, I accentuated the positive. In the long run, the humidifier would cost nothing. For equal comfort the house could be kept two degrees cooler. Fuel savings would pay for it.

The only objection to dry air I failed to mention that night— and my audience might not have accepted it as such—was that when clothes are made from synthetic fibers they are apt to cling tightly to the beautiful bodies of the girls that wear them.

What, then are the facts? Today authorities place less emphasis on what wintertime humidification contributes to comfort and health than they did years ago. The latest ASHRAE standard on the subject, No. 55-66, recommends a rh of anywhere from 20 to 60%. It cautions, however, that if it changes more than 10%, the change should not be more rapid than 20% an hour. In the past, 35% for winter was commonly recommended.

Carpets are now made—or can be treated—so they do not generate static electricity. But if they still do, touch the doorknob first with a key or even your finger nail and you will feel no shock.

A private communication from a leading consulting engineer,

who served as a Navy Captain specializing in comfort systems
during World War II, says this:

> Most humidifiers do not work after the first one or two
> years and are then forgotten, as they require maintenance.
> Many years ago the U.S. Navy required humidifiers on all
> heating systems. These became inoperative after the first few
> months and some never operated successfully. Based on
> extensive physiological studies participated in by the Harvard
> School of Health, Yale University and the University of Illi-
> nois Medical School, it was determined that low humidity,
> even 15 to 20% had no deleterious effects on the health or
> incidence of respiratory ailments among average persons. As
> a result of this report the installation of humidifying equip-
> ment aboard U. S. Naval and passenger vessels was entirely
> discontinued. The only objectors were the manufacturers of
> the humidifiers.

Some years ago, a survey was made of six large office buildings
in Philadelphia. All had year-round air-conditioning systems
designed to provide wintertime humidification. In five cases the
humidification had been turned off. Why? Because the buildings
were not built to accept high indoor humidities in the winter.
Such humidities cause the single glass windows, normally used,
to fog badly. In cold weather, they actually collect frost at night
or on the shady side of the building. Then, when the sun hits
the glass, the frost melts. Stains caused by meltage running across
the window sill and down the wall are objectionable.

But penetration and subsequent freezing of moisture can be
more damaging to exterior walls and roof than is condensation
on the windows. Table 6-5 shows the maximum relative humidities that can be maintained in a 70° room before condensation
on the outside of the glass will occur at various outdoor temperatures.

TYPE OF WINDOW	OUTSIDE TEMPERATURE, DEGREES F				
	30	20	10	0	-10
Single Glass, Wood Sash	33%	24%	18%	13%	9%
Double Glass, Wood Sash*	60%	52%	45%	40%	35%

*Thermopane, Twindow, etc.

*Table 6-5—Indoor Humidities Above Which
Condensation Occurs*

All of which helps to explain an interesting fact: The lack of the humidification function did not prevent the *fan-coil system* from becoming the most popular of the traditional comfort systems for large buildings. Nor is it interfering with the rapid acceptance of the increasingly popular new systems: the *fully decentralized* and the *decentralized heat recovery* systems (Chapter 11).

All enclosed spaces receive moisture from people who occupy them and, usually, from other sources. If such spaces could be tightly sealed, their relative humidity would constantly increase until condensation occurred or the occupants perished for lack of oxygen. Fortunately, our buildings are not all that tight. But electrically heated houses, with insulation and vapor-proofing as recommended by electric utilities, are tighter than ordinary residential construction. Result? Often, excessive indoor relative humidity in winter. Bringing in a little outdoor air will correct this condition. (See Table 6-3: 1000 cubic feet of saturated 15° outdoor air contains 0.14 pounds of water. Heat it to 75 degrees and its *relative* humidity drops from 100% to 10%.)

Houses heated with coal, oil or gas do not have this problem, even if tightly constructed. Why not? Because air is needed to burn the fuel. Hot combustion gasses going up the chimney draw into the house an equal quantity of replacement air from outdoors. Those who, in winter, sleep with their windows partly opened help to reduce indoor humidity during cold weather.

I know of only one application for which wintertime humidification is essential. That is in hospital operating rooms. But the need for it there is not for the health and comfort of patient and staff. It is to prevent the possibility of a spark from static electricity. Such a spark might cause a disastrous explosion of anesthetic gasses.

The promotion and selling of humidifiers follows a cycle closely resembling that of the seven-year locust. The first one I experienced peaked around 1933; the next one around 1940. Others followed about on schedule, with the most recent one tapering off in 1970. Why? Because it is almost impossible to design a humidifier to sell at a price people will pay and that will function properly without maintenance for, say, ten years. If it is the atomizing motor-driven spray type, the lime in the water will eventually deposit not only on working parts but also on the furniture as white dust—unless distilled water is used. If it is the evaporative type—without constant *bleeding* of excess city water to get rid of the lime— the *hardness* in the water soon fouls the equipment.

That is why there are now some humidifiers on the market with throwaway plastic evaporating surfaces. If you buy one, better take along a dozen of the replacements. Else you may not be able to find one that fits when you need it. One excellent source of indoor humidity, that doesn't lime up but may from time to time need replacement, is live plants, particularly those that like to be well watered.

Did someone give you a souvenir hygrometer, so you will know if you have too little humidity in winter, too much in summer? Forget it! I have two here in my den. Both are gifts.

Both appear to be precision instruments. Each is made by a different firm. They are 24 inches apart. On a January day, one reads 23%, the other 54% rh. I take my pick, knowing there is nothing more misleading than a hygrometer.

HEATING

The old-time radiator, placed under the window, is not all that bad, provided it is properly controlled, which it rarely is. And, with proper control, it doesn't matter if it is supplied with steam or hot water. That steam heat dries the air more than hot water is an old wives tale. Unless there is a leak, the air doesn't know what is in the radiator. But hot water heat can be more easily controlled and modulated by the individual occupant than steam heat. That is what gave rise to this fiction.

Why place the radiator or baseboard heat or warm air grille under the window? Because, as shown in table 6-3, 1000 cubic feet of, say, 40° air weighs 5.3 pounds more than 1000 cubic feet of 75° air. In short, cold air drops, warm air rises.

An ordinary glass window is not much of a barrier to the cold in winter. Cold air leaks in from outside. Warmth escapes through it to the outside. The glass itself is cold. This cools the indoor air at its surface. It drops to the floor to cause a cold draft. But not if this cold down-draft is counteracted by a warm updraft from radiator, baseboard or grille. And not—or almost not—if the window is sealed double glass. Such windows contribute greatly to indoor comfort in cold weather.

In air-conditioning equipment, heating is usually done by blowing air through the same type of fin coil as just described for cooling. Where chilled water is used for cooling, it is often the same coil. Because cool air is heavier than warm, and drops, air-conditioning grilles are often placed on the ceiling or high on side walls. In cold climates, and particularly with modern motels that have glass down to the floor, this arrangement supplies only limited winter-time comfort. It is a mistake to degrade the comfort capabilities of the heating system to accommodate cooling—especially in cold climates. But it is often done.

Increasingly, electricity is being used for heat. Here I do not refer to heat pumps. That comes later. What I refer to are resistance elements somewhat similar to those used in your electric range. This calls for a warning. The method still in general use for estimating the required capacity of a heating system ignores other heat sources: lights, people, cooking, motors, hot water heaters. This method exaggerates the heat load. The heating capacity of your furnace or boiler may also be exaggerated. Usually you will find it rated in terms of heat *input*. That means the heat contained in the coal, gas or oil it burns. But you are interested in the output, which is less—sometimes a lot less—as the equipment gets old and the heating surfaces are fouled with dust and soot, and lime is deposited in the boiler. Then, too, when your thermostat *asks* for heat there is some delay before you get it.

In contrast, electric heat is immediately and constantly (forget the Great Blackout of '65!) available at full capacity. And, for every kilowatt you buy you always get 3,412 Btu's. Therefore, the rated capacity of the electric heating equipment should be less than would be specified for a fuel-fired system. How much less? I don't know—maybe 20%, maybe 30%. If it isn't less, you are likely to have poor temperature control. The electric heat comes on so fast that, by the time your thermostat warms up and turns it off, your room is already overheated. Exception: not if you use the latest development—the higher priced SCR (for silicone controlled rectifier) heat control with a thermister instead of a thermostat.

Now just a word about radiant heat. Its use has been proposed —either gas or electric—for indoor living space. Properly done, a room might be kept at 55° instead of 75°, with a significant saving in fuel, thus heating costs. But the idea has not taken hold. However, radiant heating is extremely useful where it is not feasible to maintain the surrounding air at the comfort level. Examples: outdoor patios; shipping and receiving docks in factories.

RADIANT EFFECT

Out hunting in cold weather, have you ever stood shivering in front of an open fire? Did it consist largely of glowing coals, rather than visible flames? If so, your face and front would be roasting while your behind was freezing. Why? Because the part of your body facing the fire was receiving radiant heat. But that isn't all. The part of your body away from the fire was radiating its own heat to the cold surroundings.

No matter what the temperature level, a warmer body always loses heat to colder ones by radiation.

I sleep raw. Opposite my bed is a window wall of five sections. My bathrobe hangs on a hook by the head of the bed. Where we live, neither neighbors nor peeping toms impose the need for modesty. In winter the room is kept at 65°. I jump out from under my electric blanket. At once I shiver. Heat from my naked body is radiating through the double glass to the cold outdoors. Quickly I wrap the robe around me. Instantly the chill is gone. Why? Because the air around my body was quickly warmed to the point of comfort? No, not at all. Because the escape of radiant heat from it was instantly stopped.

If you sit at a desk near the perimeter of a glass-walled office building, close your venetian blinds or pull your drapes if you want to reduce the effect of outgoing radiant heat in winter; incoming radiant heat in summer.

In the hot Southwest you will observe air-conditioned spaces regularly maintained at 70° or less. They say they need it for comfort. And they do. Why? Because the uninsulated walls of the buildings are many degrees above room temperature. In winter the reverse condition exists. The only way to compensate for it is by a decrease or increase in air temperature.

How much? For every degree the average temperature of the surrounding walls is below room temperature raise the room temperature one degree. And, vice-versa, where the wall temperature is above room temperature. But, if you have a lot of unshaded glass, you may want to go even further.

There is, of course, one other way to reduce radiant effect—the best way! That is by the use of wall insulation. Unfortunately, the decision whether or not to use insulation is normally based on economic factors, not comfort. Will the resulting reduction in the cost of heating and cooling pay for it?

THIS DECIBEL BUSINESS

Killick-Nixon were selling more noisy $475 prewar room coolers than anyone else in sweltering Bombay. Elsewhere, customers had complained of the noise. Killick's, it seems, had solved the problem. How? In the corner of their large display floor was a small glass-walled *closing* room. High on a wall pedestal stood the usual—for Bombay—16-inch oscillating electric fan operating on HIGH. Also installed was the demonstration room cooler—operating.

The prospect would enter the delightfully cool closing room. In due course, the sale would be made. Then the salesman would say, in effect: "Look, we want your wife to be pleased too. You want to make sure she won't find the unit too noisy for your bedroom. Why not bring her in to listen to this one here, before we install yours?" A married prospect at once recognizes the soundness of such advice.

The wife would arrive. The benefits of air-conditioning would lose nothing in the telling—"prevents prickly heat, protects your complexion." Then would come the clincher. "This is an unusually quiet machine. But your husband wants to make sure it won't disturb your sleep. So he would like you to listen to it without any other sound in the room." Then, just as he had done when selling the husband, the fan—sounding like a Piper Cub about to take off—would be stopped. The air-conditioner could scarcely be heard. The lady was delighted. And said so. How, thereafter, could she possibly complain about noise?

When a loud noise is suddenly stopped the surrounding space seems much quieter than it did before the noise started. Starting and stopping of fans and compressors can be more annoying than if they operate continuously at full noise level. You will soon become accustomed to it, provided it is below the speech interference level. An acceptable sound level is just as important for human comfort as temperature, humidity and air motion.

Sound is measured in decibels (db). But you will not react the same to two sounds, both of, say, 70 db. One may be music, the other an excruciating racket. It depends on frequency distribution. Decibels are peculiar things. Say you normally have a sound of 60 db in a room—street noise and conversation. You turn on an air-conditioner that generates 60 db—that is a quiet one! Do you get 120 db in the room? No, only 63 db. What, then, is the level of various sounds you are regularly exposed to?[11]

A quiet room	40 db
Normal speech	60 db
A loud voice	70 db
The loudest yell	90 db

Sound reaches you as a vibration transmitted, usually, through air. It has certain frequencies. Middle C is 264 cps (cycles per second). Most speech is between 100 and 300 cps. The most critical noises are in the range of 600 to 2400 cps—the second D to third C above Middle C. When we are children our ears can pick up everything from 20 to 20,000 cps. By age fifty the upper limit has dropped to 10,000 cps—still amazing. The range of sound magnitude the human ear can hear is even more amazing. In a soundproofed room your ear can hear a sound of 1000 cps that has only one millionth of the intensity of the sounds to which you are regularly exposed.[12]

At 30 cycles per second you can not hear 60 decibels. At 1000 cps 60 db would be loud.[11] So, when someone talks about 60 db it doesn't mean a thing. Yet I just said that a 60 db air-conditioner is a quiet one. That is because, in practical use, three different scales are used: dbA, dbB and dbC. Scale dbC represents the true sound level. Sound meters that read in dbA have been modified to respond as does the human ear. It's the dbA scale that is used for air-conditioning systems. So, when a layman talks to you knowingly about decibels, let him finish. Then, if you enjoy one-upmanship, use Stephen Potter's favorite ploy: "Yes, but not on the A scale."

When an engineer writes a sound specification, he may simply say that the sound level should satisfy NC-35. By this he means noise criteria curve No. 35 as published by ASHRAE. Each curve is made up of a different decibel rating for each of eight frequency bands. For NC-35 it looks like this:

CYCLES Per Second	DECIBELS	CYCLES Per Second	DECIBELS
20 - 75	63	600 - 1200	37
75 - 150	55	1200 - 2400	35
150 - 300	47	2400 - 4800	33
300 - 600	41	4800 - 10000	32

But this is of no use to you unless you are buying an air-conditioning system and can get a valid sound level guarantee from the seller. In that case, if you can afford the cost, insist on levels that do not exceed these:

APPLICATION	NC CURVE
Churches, Movie Houses	25 - 30
Private Residences	20 - 35
Hotel rooms, Apartments, Private Offices	30 - 40
General Offices, Restaurants, Stores	35 - 50

Table 6-6—Recommended Sound Levels[12]

Use minimum values in the country and other quiet locations; maximums are OK in noisy ones. And remember, anything below NC-30 is *very* hard to come by.

The same equipment may sound much louder in one room than in another of the same size, even though the sound entering from outside and that generated inside are each the same. Acoustically, one room is *hard,* the other *soft.* A typical hard room has bare plastered walls and ceilings, wooden or masonry floor, undraped windows. An acoustic ceiling, carpet, drapes and tapestry wall hangings will make it soft and reduce the indoor sound level.

But you should be concerned with more than indoor sound level. Increasingly, the sound made outdoors by air-conditioning equipment is coming under fire. More and more cities and states are enacting codes to control it. Are you planning to install equipment that will add to the outdoor sound level? If so, better make sure that it satisfies your local code. And, if you have none, ask ARI for a copy of their model code and make sure you satisfy it. This would give you a strong defense against legal action by irate neighbors or your local community—once it adopts a code—if either objects to the outdoor sound level of your equipment.

Is it possible for an occupied space to be too quiet? Anyone who has stood quietly in a soundproof room hearing his heart go thump . . . thump . . . thump knows the answers is *yes*. So does the man who says *gesundheit* when his neighbor in the next apartment sneezes. Below a certain sound level the creaking of pipes and ducts, as they expand and contract from temperature changes, ticking of clocks, chirping of crickets, a barking dog, starting of an automobile, all become increasingly objectionable as room noise level is reduced. So, does snoring.

That is why, according to the Associated Press, Ronald Carr of Melbourne, Australia, invented a device that would let his wife sleep, despite his snoring. It is a sound maker she keeps under the bed. Its volume can be adjusted to suit the occasion. "It doesn't stop him from snoring. Let's get that straight from the start. But it does allow me to put up with it", says Mrs. Carr. And a well-known specialty shop in New York City sells a device called Sleep Sound for $19.50 plus tax. Allegedly it *whooshes like the breeze in the trees.*

Chapter 7

It Was A Long Time Coming

Swathed in skins of animals he had killed for food, Og was hunkered inside his cave, near the foot of a hill in the Neander Valley between Düsseldorf and Wuppertal, in Germany. In front of him were the glowing coals of a fire. He had made it himself. Without it he could not have survived the wintertime rigors of the Ice Age during which he lived.

Yes, Og could with difficulty build a fire. He knew how to bank it with ashes, so the glowing coals would survive while he was out on the hunt or sleeping in his cave. When it was that Og's ancestors first learned to make and, within limits, to control fire we do not know. But we do know that our Neanderthal ancestor was able to do it. And we rather accurately know when he lived—seventy thousand years ago, give or take perhaps thirty thousand.

ESCAPE FROM THE COLD

The means of creating warmth was mankind's first step in its efforts to escape the debilitating extremes of nature. Thousands of years elapsed before the art of heating escaped the limitations of the open fire. But we know that ancient Romans of wealth ingeniously engineered ventilation and panel heating into their baths. And it was in 1742 that our most versatile genius of revolutionary times, Ben Franklin, invented that great improvement over the open fireplace, the Franklin Stove. It gave increased heat and safety with economy of fuel.

In 1815, Robertson Buchanan, a Scottish engineer, published a book on heating and ventilating. From then on progress in this field accelerated. By the end of the century a variety of space heating systems was in general use. The cook stove warmed the kitchen. The potbellied parlor stove served the living room and, by means of a ceiling register or stovepipe extension, sent some of its benign warmth to the bedrooms above. Soon gravity warm air furnaces and radiators, supplied by hot water and steam from a central boiler, came into use for private homes, having first been used in larger structures. The latter were also served, then, with forced air systems, the air being heated from a central boiler by banks of steam coils.

The space heaters of 70 years ago were hand-fired with wood, coal and coke. Small, smelly, portable heaters using *coal oil* were also available. Soon chains and pulleys were devised with which furnace and boiler dampers could be adjusted by hand, without going into the cellar. This crude arrangement was then followed by primitive controls, actuated by a thermostat. Automatic heat, by means of coal stokers, and oil burners, was first introduced in large central heating plants. By the 1930's it was the latest thing for the modern home. Then came natural gas and, more recently, electric heat. Today, the use of coal for space heating is almost *obsolete,* oil is on the *decline,* natural gas is most *popular* and the use of electricity for heating is growing most rapidly in new construction.

REFUGE FROM THE HEAT

Lost in antiquity is the origin of an ancient practice observed in the hot, dry plains of central and western India. There, loosely woven mats are placed in wall openings that face the prevailing winds. A servant sprays the mats with water often enough to keep them constantly wet. The breezes passing through the mats are cooled by evaporation and provide the master with some relief from the heat.

There are arid parts of the United States and elsewhere in which comfort cooling is still accomplished simply by the evaporation of water in a stream of outdoor air. Elsewhere, refrigeration—mechanical or by the use of ice—is necessary. As early as 1000 B.C., a Chinese poet told how, in the second month of the year, the ice is harvested. In the third month it is put in storage and in the fourth the ice house is opened "early in the morning", so the ice may be used. During the time of its splendor, in 775 A.D., so it is said, Caliph Mahdi of ancient Baghdad had slaves bring snow from the mountains for his do-it-yourself air-conditioning project. The clever Caliph had his palace built with double walls. The snow was packed in the space between. Perhaps the Caliph had heard of a more primitive effort of the same kind, made in the third century A.D. by the Roman Emperor Heliogabalus.

Real progress had to wait until 1748. That year, William Cullen, of Glasgow, demonstrated the production of cold by evaporating ether in a partial vacuum. A forerunner of today's refrigeration machine was proposed in 1805 by Oliver Evans of Philadelphia. But Dr. John Gorrie (1803-1855) gets the credit for building the first machine that actually cooled people.

As Director of the U. S. Marine Hospital in Apalachicola, Florida, Gorrie had a contract with the U. S. Government to care for sick and injured sailors. In 1833, Gorrie hung buckets of ice in hospital rooms, then blew air over the ice to cool the room for malaria and yellow fever patients.

The retail price of Boston lake ice was then a whopping $200 per ton along the Gulf Coast—when available. Lack of ice with which to relieve his patients was the incentive that pushed Gorrie to develop his machine.

Under a pen name, Gorrie first published his ideas in 1844. As he may have expected, they were ridiculed by the press. But a Boston financier heard of them, was convinced, and put up the necessary cash. Gorrie had the machine built in a New Orleans shop. It was a success. In 1849 it made ice and it cooled air.

In 1850 Gorrie applied for a U. S. patent. It was granted in 1851—No. 8080. His machine employed the simplest of cycles. It compressed air which, as anyone who has pumped up a tire knows, heated it. This heat (i.e., energy) was removed by passing the hot air through pipes submerged in water. Then the compressed air gave up more energy by being used to operate what was essentially a steam engine running on compressed air. This caused the temperature of the air exhausted from the engine to drop. Actually, it was cold enough to produce ice![15]

Later, an improved version of this inefficient method was extensively used on U. S. Naval ships. Today it is still in use. In highly efficient form it provides comfort for the passengers of the Jumbo Jet—the Boeing 747.

The efficient ammonia compression refrigeration system got its start in 1873, when an emigrant from Scotland, David Boyle of Mobile, Alabama, operated his first machine in Jefferson, Texas. It is still the most popular type for large-scale manufacture of ice, cold storage and food freezing. But, for safety reasons, its use for comfort cooling has always been limited. At about the same time a similar system, using carbon dioxide, was developed. Despite its low efficiency and high operating pressures, it had two virtues. CO_2 was cheap and relatively safe. Prior to World War II it was extensively used for shipboard refrigeration and air-conditioning, and for theatre cooling.

Willis H. Carrier demonstrated his first centrifugal refrigeration machine in 1922. Initial installations were for industrial use. It used a nonhazardous, low pressure refrigerant. For human comfort, it was first put in operation on Memorial Day of 1925 in New York's Rivoli Theater. This type of equipment gradually superseded CO_2 systems for theaters and other large comfort installations. The smallest centrifugal to prove feasible was large enough to cool a theater of 1000 seats. Thus, it was not practical for small jobs.

In 1930 a long leap forward was made for adding coolth, the complement of warmth, to means for gaining year-round personal comfort. It was the announcement of a new, safe, relatively low pressure refrigerant—the halogenated hydrocarbon named Freon 12 by its first producer, DuPont. It was particularly useful in filling the need for safe systems with capacities below those for which centrifugal machines were suitable.[15]

IN THE BEGINNING

Air-conditioning for human comfort was born in 1919—the offspring of a marriage of convenience. Conceived by the Profit Motive, its proud parents were an unnamed central heating plant and a Wittenmeyer CO_2 refrigeration machine. Its midwife was the theater chain of Balaban and Katz. And the child's foreordained career was to fill movie houses during hot weather.

Thus, appropriately, it was born in Chicago's Central Park Theater.

True, a few stillbirths had occurred prior to 1919. About 1870, a primitive form of mechanical refrigeration was used to cool an Indian Rajah's palace. In 1893, a professor asked in *North American Review:* "If they can cool dead hogs in Chicago why not live bulls and bears in the New York Stock Exchange?" In 1902 it was done. The system functioned 20 years. Then, in 1905, Walter L. Fleisher used the refrigerating machine that made ice for the skating rink of the New York Hippodrome also to cool air for comfort. In 1906 my friend, Arnold Goelz, designed a system for the Pompeiian Room, Congress Hotel, Chicago, which was updated in 1911. Carrier first tried its hand at comfort cooling by doing the dining room of Milwaukee's Wisconsin Hotel in 1913, as already reported.

But all this was prelude. These isolated instances did not become a continuous flow—not until Balaban and Katz, and others, gave theater cooling their uninterrupted support. In 1920 Fred Wittenmeyer supplied B&K with a second system for Chicago's Riviera Theater. Then faster and faster they came. Soon the steady stream was building an industry. By the early 1930's —despite The Great Depression—mechanical cooling was standard for all deluxe cinema palaces that were springing up in our larger cities and, increasingly, abroad. Soon cooling installations were also being made in department stores, restaurants, railway dining cars and other business places that wanted to attract the public in summer.

COOLING—BY THE TON

The amount of cooling required to provide comfort in a given space during extreme summer weather is much greater than, in the early days, common sense indicated it should be. Example: back in 1929 the virtually new State Theater in Sydney, Australia, decided to install its second air-conditioning system. What was wrong with the first one? In the basement there was a box large enough to hold six blocks of ice. A blower forced air over the ice and into the theater. A trap door in the sidewalk enabled the box to be loaded. This was done once a day, from a truck. The standard ice block weighs 300 lbs. So, full, the box contained 1800 lbs. Mama Mia that's a lot of ice! But is it, really?

When estimating the cooling load it is done in Btu per hour. The cooling power of refrigerating machines is, however, often expressed in tons: "I have a 1-ton unit in my office." "The man said a 3-ton machine is large enough for my house." "They say it's going to take 900-tons to cool the new 30-story office building." So what does it mean, a ton? Is it the weight of the machine? No, it is its cooling power or capacity.

For theaters, the rule of thumb is 15 to 20 seats per ton. For a 2000-seat house that means, say, 120 tons. We do not know whether 1800 lbs. of ice is enough until we translate tons of ice into tons of cooling power, as produced by the refrigerating machine.

It takes heat to melt ice. Ice melts at 32° F. The heat it takes to convert one pound of ice into one pound of water, at 32°, is 144 Btu's. It is called the *latent heat of fusion*. And that 144 Btu's is the amount of cooling it will do. A ton of ice weighs 2000 pounds. Thus, when it melts, how much cooling does it do? Obviously 2000 times 144 or 288,000 Btu.

A ton of refrigerating machine capacity is, therefore, 288,000 Btu of cooling done over a period of one day—24 hours. So, if you divide 288,000 by 24 you get 12,000 Btu of cooling per hour. That is what is meant by a ton of refrigeration.

Sometimes you hear *ton* and *horsepower* used interchangeably. This is because a refrigeration compressor used for air-conditioning often requires a motor of one hp for every ton of its cooling capacity. Under extreme conditions, it takes more. Under favorable conditions, it is less. In either case it does not include power for fans and pumps. It is a misleading practice that both ARI and AHAM (Association of Home Appliance Manufacturers) have long opposed.

Now, let's revert to the State Theater. It needed 120 tons. In terms of Btuh this is 120 times 12,000 or 1,440,000 Btuh.

What will the 1800 lbs. of ice do? Well, 1800 times 144 amounts to a total cooling power of 259,200 Btu. How many hours will that cool the theater? To find out divide 259,200 by 1.44 million. The answer? 0.18 hours. That is 10.8 minutes, if their contraption could have melted the ice that fast, which it couldn't. So, from the standpoint of cooling that theater, 1800 lbs. of ice a day wouldn't make a pea hole in the snow! No wonder they ordered a second system. I was there, watching Al Jolson in the Jazz Singer, the first night it was used. It did the job!

THE UNSUNG HERO

Here is an item from the Syracuse Herald American for Sunday, June 20, 1965:

600 REFRIGERATORS SET FOR PIONEER HOMES

The Syracuse Housing Authority has ordered 600 new refrigerators to be installed in the Pioneer Homes public housing development, William L. McGarry, executive director, reported last night.

McGarry said the appliances will replace those originally installed when the housing project was opened in 1940 for tenants who desire them.

Is a 25-year electric refrigerator life unusual? Not at all. The average life of a domestic refrigerator is conservatively considered to be 15 years. But what has this to do with air-conditioning? I am coming to that.

The heart of both an electric refrigerator and an air-conditioning system is the compressor—its most costly component. On the average it operates in a refrigerator at full speed almost half the time—4000 hours out of 8760 in a year. In 15 years that is 60,000 hours. So what? So if your automobile operated 60,000 hours at only 50 miles per hour that would be three million—repeat million—miles. And that is 120 times around the world, at the equator, or six round trips to the moon. And it does it without a valve job, new points, new plugs or even an oil change. In fact, unless the machine is defective, neither lubricating oil nor refrigerant are added over the 15 or 20 or even 30 years during which the refrigerator may have operated. Search your mind. Can you find any other electro-mechanical device with a comparable record of reliability and operating life?

What accounts for the long life of the electric refrigerator? In terms of the service it renders the public, it is the most important new refrigeration development since the ammonia refrigeration system was introduced 100 years ago. I refer to the *hermetically* sealed refrigeration circuit ("Her-met-ic, adj. 1. made air-tight by fusion or sealing", pg. 665, The Random House Dictionary.)

Why this accent on hermetic? Because, while Achilles had his heel, the refrigeration compressor had its seal. This was always its weak spot, particularly in the smaller sizes. A compressor has a shaft that must be rotated by a motor, an engine or other prime mover. For many years that shaft projected from inside of the compressor to the outside. It connected the compressor to its source of power.

The inside of a compressor is full of refrigerant gas—sulfur dioxide, ammonia, carbon dioxide, methyl chloride or one of the Freons, usually Nos. 11, 12 or 22. Usually the refrigerant is under pressure. In a domestic refrigerator the loss of as little as half an ounce may cause the machine to malfunction. Thus, where the shaft leaves the compressor, some form of seal is interposed to prevent leakage. (Sometimes the refrigerant is under a vacuum. Inward leakage of air will likewise cause malfunctioning.) Seal leakage is as old as the refrigeration industry. Over the years seals have been greatly improved. But if your automobile air-conditioner ever needed gas, chances are it was due to seal leakage.

The best solution to the seal leak problem is to eliminate the seal entirely.

This was first accomplished in the Audiffren-Singrun machine, developed in France and then, in 1911, also made in the U.S. by General Electric. External power rotated the entire machine. The crank shaft, all inside, was counterweighted and stood still. Then a Swiss firm, Escher-Wiess, went a long step further. An alternating current motor consists basically of two parts, a rotor and a stator. The Swiss mounted the rotor on the end of the compressor shaft and sealed the entire assembly inside of a steel shell. The stator, separated from the rotor by a thin wall of steel, was on the outside.

By 1926 General Electric had the right answer. That year its Monitor Top Home Refrigerator was placed on the market.

It had the entire driving motor inside of the compressor shell, cooled by the refrigerant—sulfur dioxide, initially. Until then this market had been dominated by Copeland, Frigidaire, Kelvinator and others. GE's hermetic design made their so-called *open* machines obsolete.

By World War II it was obvious that the hermetic compressor could do for air-conditioning what, by then, it had done for household refrigerators. By 1949 all postwar room air-conditioners employed true hermetic compressors or those miscalled semi-hermetics. The true hermetic circuit has its motor-compressor assembly *welded* into a steel shell with the rest of the all-welded refrigerant circuit welded to it. So called semi-hermetics are better called accessible hermetics. Usually the parts are *bolted* together.

Welded hermetic compressors for air-conditioning are now produced in sizes as large as 11 tons with larger sizes being developed. Accessible hermetics with reciprocating compressors go to well over 100 tons. Those with centrifugal compressors up to 1000 tons or more.

When properly made and applied in an environment conducive to long life, there is no reason why the truly hermetic compressors, as used in air-conditioners, should not operate as long as they do in refrigerators. Although that ideal has not yet been achieved, it has been approached. Note, too, that for comfort cooling the compressor does not operate 4000 hours per year. It almost always operates less than 2000 hours a year in the United States and often less than 1000. Obviously, therefore, it should last many years longer in an air-conditioner than in a domestic refrigerator.

To the welded hermetic compressor—taken for granted by its

user; derisively called the *tin can* by those who do not use it—belongs the credit for the amazingly rapid acceptance of *small* air-conditioning since World War II.

THE PROOF OF THE PUDDING

As used here, *air-conditioning* means year around indoor environmental control. It does *not* mean simply cooling. Nor does it mean heating. It means whichever and whatever is needed to provide physical comfort for the occupant of an enclosed space. And, however it's done, whatever is used to do it with, the worldwide acceptance of air-conditioning has been phenomenal.

Willis H. Carrier, the leading pioneer of the industry, started designing air-conditioning systems, marketed under the Carrier name, in 1908. Initially, his systems were exclusively designed for industrial processing. Not until 1913 did he design a comfort system. It was for the dining room of Milwaukee's Wisconsin Hotel. By 1949 Carrier Corporation sales amounted to $46 million, with comfort systems dominant. By 1969 sales had multiplied almost twelve times, to $536 million, mostly by internal growth, not acquisitions. And, by far, the major portion of this was for human comfort.

Between 1948 and 1968 the total value of air-conditioning equipment installed in the United States (exclusive of transportation equipment) multiplied eleven times, from $400 million to $4,400 million. Someone must like it. Whether for good or for bad, Las Vegas, for example, would be impossible without summer cooling. Even in his time, F. Scott Fitzgerald could say, "There is much less weather than when I was a boy."

True, a well-known male opera star recently stopped in the middle of a performance in Miami and refused to continue until the air-conditioning was shut off. But that is the privilege of a prima donna. Then again, the booking agents for Lawrence Welk and Frank Sinatra refused to let their stars play in Rochester, New York's War Memorial during the summer of 1969. Was it because Lawrence and Frank insisted on cool comfort? Or the lack thereof? No, it was because, since the building was not air-conditioned, the audience would stay away in droves.

However it is done, the purpose of air-conditioning is to help us escape from the cold and have a refuge from the heat. But, for all we know now, summer cooling may have a beneficial effect on human intelligence. At any rate, the Journal of Experimental Psychology has reported that baby rats born and raised in a cold room were smarter than those in hot rooms. And 2300 years ago, the Greek physician Hippocrates observed that in hot, humid lands, "the inhabitants are fleshy, ill-articulated, moist, lazy and generally cowardly in character. But where a land is bare, waterless, rough, oppressed by winter's storms and burnt by the sun, then you will see men who are hard, lean, hairy; of more than average sharpness and intelligence in the artisan and in war of more than average courage."

Many years ago I had occasion to write: "Extensive travels in the tropics over a period of 15 years have convinced me that the enervating effect of the climate is the key reason why this potentially most productive part of the earth is, actually, the least productive inhabited area. The minimal effective antidote to the tropic's enervating effect on humans is an indoor climate conducive to a good night's rest. This means some form of bedroom cooling."

And in the same vein, the late great Roger W. Babson has said: "Air-conditioning should do for the tropics what coal has done for the northern countries. . . My chief interest in air-conditioning is in connection with the utilization of the tropics to their fullest extent. Comparatively small areas in Brazil could absorb millions of people and give them a high standard of living. Air-conditioning can open up, freely, great frontiers. . ."[16]

Then, too, there is the universal appeal made by Beardsell and Company in an advertisement in their local paper, in tropical Madras, India: "Make love in comfort. Get a room air-conditioner NOW!"

Yes, warmth and coolth, as desired, should make us completely unaware of our indoor environment. Between them, they can free our minds for more productive matters. That is the implied promise of air-conditioning. And that is the promise that Tradition has made it so difficult to keep. Nowhere have the forces of Tradition been more powerful than in the industry that provides our indoor environment—the building industry.

This is the industry of which the comfort business is a part. This, then, is the likely suspect on which to serve the indictment that was drawn up in Chapter 1.

Chapter 8

The Industry That Failed
To Keep Its Promise

The fields of science, medicine and aeronautics are known for their rapid progress; the building industry for its untiring efforts to stay the same. It is still in a primitive stage, the craft stage, the hand-hewn stage of technical development. It still lays one brick on the top of another, as was done in ancient Egypt in the time of Moses. The holding back of what is inevitable is its dominant characteristic.

Actually, the building industry is no industry at all. It is a poorly coordinated conglomeration of craft unions, manufacturers, money lenders, politicians and government agencies. Over the years they have fought through rulings, codes, gentlemen's agreements, make-work and other restrictions that stifle progress.

The air-conditioning industry is a part of the building industry. In comparison with some segments of that industry, air-conditioning has made great progress. In comparison with many areas of science and technology, it has not.

As we have seen, manufactured discomfort is everywhere. Equipment and systems are installed that satisfy an enigma of instruments but not the people they are supposed to serve. The last thing with which the industry is concerned is, too often, the comfort of the occupant—not just physically, but also financially. It is a peculiar industry: it avoids talk of what it has to sell—the physical well-being of the building occupant; it accepts customer dissatisfaction as normal. To the extent that everybody likes comfort but not air-conditioning, the comfort industry has so far failed to keep its promise.

THE OWNER

The owner is properly responsible for what goes into his building. This is true, irrespective of the size of the structure. It is true, when one hot July day, he stops on impulse on his way home to buy a window unit for his bedroom, complete with *qwik-kit* for do-it-yourself installation. It was true when the Port of New York Authority approved the largest system so far produced for its new World Trade Center.

But the owner must in all cases depend on others. If for no other reason, it is because a bewildering array of choices persist on the market, since "you can sell anything if you go about it right." The man who buys the window unit depends on the integrity of its maker and the veracity of the dealer who sold it. As the size of the job increases, so do the number of influences involved. Consider a large new office building. Who else, besides the owner—who, too often fails to assume the fatigue of investigating for himself, adopts a completely hands-off attitude and, in most cases, is a novice in this area—influences the design of the air-conditioning system?

It is obvious that this includes the owner's professional advisers: the architect and the consulting engineer. Of course it includes one or more contractors and local building codes and fire regulations. But it also includes the mortgage holder, one or two local public utilities and, in the case of office buildings, the principal tenant—after whom the building may be named. And there is one other powerful factor that is always in the act: money. How much will it cost? Will it be within our budget? And, once it's installed, how will it be operated? That, too, involves money.

Early in May, I spent two nights at Boston's Statler Hilton. The guest room comfort system was one designed to give—and was capable of giving—each guest a choice of either heating or cooling under his own control. After a day of meetings I returned in the late afternoon. It was sunny. My room faced west. It was hot. I turned the controls to *cooling*. Nothing happened.

Having once worked out of the adjoining office building, I knew the place. No one was in the engine room when I got there. Going to the gauge board I observed the chilled water thermometer. It should have read 45° F. It actually read 78°. "Why," I asked the engineer when I found him, "can't I have cooling in my room?" "You see, sir," he said, "we are not allowed to turn on the cooling before May 15th." Weather Bureau records show Boston temperatures as high as 88° in April and 93° in May.

The night of October 11, 1970, was chilly in Detroit. It was after eleven when I checked in at Stouffer's Northland Inn. One look told me the *3-pipe* guest room comfort system had been designed to give me a choice of heating or cooling. I set mine for HEAT. None came. Just for fun I then tried COOL. Ditto! The system was off. To keep me warm, it took two blankets and the bedspread.

The frugality of two leading hotel chains was showing.

It was a hot night in Washington. The high-rise Holiday Inn was only a year old. Its facilities were excellent. Even the guest room comfort system was beyond reproach. It was June and I was hot. Only a feeble stream of cool air was dribbling lazily out of the supply grille. It wouldn't have cooled a phone booth. After a few tries, the panel that gave access to the air filter was removed. You couldn't see the filter—just a felt-like mass of lint and fibers shed by the new carpet over the past year. The manager was called, the filter cleaned, refreshingly cool comfort soon was restored.

All of which is just another way of saying that many systems that have the *capability* of doing it are not providing full-time comfort because of the way in which they are operated and maintained. When was the last time you cleaned or changed the air filter on your room cooler, warm air furnace or residential air-conditioning system?

Let us now revert to those responsible for the design of the system. Why? Because often it lacks the capability of providing satisfactory comfort conditions, no matter how well it is operated and maintained.

THE ARCHITECT

Russ Jacobsen (not his real name) is Chief Engineer of the Mortgage Department and an officer of one of America's largest insurance companies. He was looking over the plans of a new 30-story office building for downtown Boston. He didn't like what he saw. "Look," he said to the architect, "you have the glass right down to the floor. We won't loan the owner the extra money it takes for double glass. With single glass down that far, no one will be able to sit within six or eight feet of the outside walls on a cold day. Why all the glass?"

Glass, unless double, is the hero of the architect; the enemy of comfort. So what was the architect's reply? "It gives the building bouyancy." After Russ told me this he added, puzzled: "How does an engineer answer that?" "Don't try," I said and quoted the opening sentence of a talk I had recently given before a national convention of architects: "There are two kinds of people I will never understand. One is women, the other is architects." (It got their attention.)

Another time, two of us were in the New York offices of one of the world's greatest architectural firms. We were talking to the senior partner in charge. They were designing the new home office building for a large New York bank. The bank was unhappy with the air-conditioning system in its existing building. Since it was our bank, its top officer had arranged the meeting. He could see the advantages of one of the new systems—one that would provide more comfort for less money.

After polite introductions and the usual sparring around, we got down to cases. No, they did not do their own engineering in New York. They had farmed out the mechanical design to Harris, Tree and Cotton. This leading firm of engineers was well-known to us. We were sure they would propose the exact same system (method, not make) that was giving unhappy results in the bank's present building. So, gradually, we made our pitch, feeling our way as we went.

Why not, we suggested, let the engineers lay out both systems, assuming they were reimbursed for the extra work? Why not, then, estimate the comparative costs of the two—both first cost and operating cost? Then, why not present the pros and cons of each system to the owner—particularly their respective comfort capabilities—and let him decide for himself?

Our reception had been warm and friendly. Suddenly we noticed a chill in the air. Straightening up to his full six-feet-two, the senior partner made his position clear: "We are the architects. We, not the owner, are designing the building. What the engineers propose that is what we will put in." Case dismissed! And that, eventually, is just what happened. The bank was saddled with a $900,000 penalty in first cost plus a penalty in operating cost conservatively estimated to be $58,000 per year, for part-time, not full-time comfort. It was the bank's money, not the architects'.

An Ivy League university was building a new $34,000,000 Engineering Quadrangle. The architects were another nationally known firm. The design was completed, bids were requested, the contracts were placed. The mechanical contractor ordered the heating and cooling equipment, as specified. Its maker had long since given the design engineers complete technical data. This included the minimum size of wall openings for outside air.

Acting as if the engineers, not the Good Lord, had created the laws of nature and thus could easily change them, the partner in charge defied his engineers and arbitrarily reduced the size of these openings by 29%. After thirteen months of futile negotiation, their size remained unchanged. The equipment was urgently needed on the job. The manufacturer was left with two choices: Cancel the order and lose the hard-won goodwill of the architect, engineers, contractor and university or modify the equipment. That is what he did. He increased motor size, speeded up the blowers and forced the air through the undersized openings. Results? For the life of the system, slightly higher operating cost, but objectionably higher noise level.

Some architects do have a keen appreciation of matters technical. They understand and accept technological progress and accommodate their building designs to it. But they are in the minority. Many of them—monument-oriented—proceed with only the aesthetic creation of a new structure in mind. Initially, they ignore the electrical, mechanical and sanitary facilities without which it would be an empty, uninhabitable shell. Later, having by then allotted too much of the available budget on what has already been done, these essential elements suffer. They are modified from what they should have been, to match what has already been done, using what is left, if anything, in the budget. Too often this is to the detriment of both the owner and, in the case of air-conditioning, the comfort of his future occupants.

This helps to account for the trend, on the part of experienced, repetitive builders and developers, increasingly to employ and control their own in-house architects.

THE CONSULTING ENGINEER

A few years ago two young graduate engineers left their home in France and came to the USA to learn all about air-conditioning. This was understandable. Their fathers owned and managed one of the largest mechanical contracting firms in France. It had branches in key cities of that country as well as in Belgium, Algiers and Morocco. After attending factory training courses at Trane, Honeywell, York and Remington, they both went to work for a leading Philadelphia consulting engineer. This is where they hoped to try out what they had so far learned.

Now, it happens that there are more ways to air-condition the kind of building for which consulting engineers are employed than ways to cook an egg. A system that may be *best* for one structure may be poorly suited to another. Thus, before a final decision is made, experienced judgment must be used to select those three or four systems which appear to best suit a specific application. Then, detailed studies of each system must be

carried forward. The object? To find the optimum combination
of first cost, operating cost, space requirements and—too often
near the bottom of the list—comfort capability.

At least this, with true French logic, is what these bright young
engineers properly expected. They were in for a shock. The boss
picked up a set of architectural drawings he had just received.
He gave them a quick glance. Then he handed them to Jean—or
was it Pierre?—and said: "For this one, figure a two-pipe induction system, like the one that went into the Blank Building on
Broad Street."

Had this been Los Angeles there is a better than even chance
he would have said, "figure a dual-duct system". In Minneapolis
it might have been "a 3-pipe fan-coil system." The point is, at
least to a degree—a declining degree, fortunately—that there
are geographical styles to air-conditioning. There is a tendency
to perpetuate the system that, in each area, was most aggressively
promoted or first introduced there.

Most disturbing, because it is more prevalent, is what was
said about another large firm of architects/engineers. A successful mechanical contractor was talking about them in my presence. "I could bid a Major, Noah, Crabtree and Jester job by
the square foot, without seeing the plans. They are all alike."
The design engineer—whether an independent consultant or the
head of the architect's engineering department—is the most
important influence in the selection and design of any air-conditioning system. Many design engineers consistently tend to
favor the same system. If you want a different system, the most
painless way to get it is to engage a different engineer—having
found out first what he favors. It is easy to see why this is so.

The industry started with large buildings, not private residences. It started with central systems, not room coolers. Each

basic system concept was initially put forth by a certain manufacturer. He was the one who had conceived, developed and possibly patented it. He produced the often unique equipment required for his particular system. Initially, his own men designed each job.

Many of these designers left to become independent consulting engineers. Other consulting engineers were actually trained at factory seminars, or in their own offices, on how to design the particular system promoted by one or another manufacturer. Then, as they grew older and larger, they passed this knowledge on to their associates.

Such men are naturally tied to the specific products and systems of the manufacturer who trained them, by an intellectual umbilical cord they hesitate to break. Then, too, it is easy and takes far less man hours to hand a set of prints to Joe, over in the corner, and say, "Here is one like the Main Street Building we did back in 1961. Get out the file and adapt it to this job." Studying all likely alternatives, before deciding on which system to use, takes much more time, costs much more money. Yet the fee stays the same.

Most consulting engineers deserve great respect and admiration for their ethics and competence. There are some, however, also professionally licensed and active as air-conditioning designers, who will *gold plate* the job where the budget permits, since the higher the cost the more the fee. Others have regrettably limited competence. One way they keep going is to depend on the favored local representative of a manufacturer to do their design work for them. The representative naturally specifies the equipment he sells—*ties up the specs,* in the parlance of the trade. The recipient of the *free* engineering then charges his client the full fee or a *special price*—in either case unearned.

Irrespective of who does it, every system of air conditioning is designed by or under the direction of one man. The first step is to determine *the load*. By this is meant the capacity of the heating equipment needed in winter; the cooling equipment needed in summer, to provide assumed comfort conditions. It depends on the size, design, occupancy and geographical location of the building. Typical *design conditions* for St. Louis, for example, would be 70° F indoors and 2° F outdoors, to determine the necessary heating capacity. For cooling, the indoor temperature might be taken at 75° with 98° outdoors and since "it isn't the heat but the humidity", the removal of 50% of the moisture in the outdoor air, under design conditions, would also be assumed.

Competent engineers agree closely on the capacities required for a specific job. The procedure for determining them is well standardized and fully covered in the 4-volume ASHRAE Guide. The weak link has to do with the economic and comfort consequences of system selection. No longer is the design of the comfort system primarily an engineering problem. Engineers must increasingly think in terms of *most comfort per dollar*. What must the designer know, to do this effectively? He should, in the case of each specific project, be able to give reassuring answers to these questions:

1. With which systems, in addition to the one he is proposing, has he had actual experience for the same type of structure?

2. What are the comfort capabilities of the proposed system, *when operated as it is likely to be operated?*

3. For which systems, in addition to the one he is proposing, does the designer have comparative data as to first and operating costs?

4. What is the minimum permissible period within which the

cost of the equipment may be written off for tax purposes? (Depending on the system, IRS regulations provide for ten, fifteen or twenty-year periods.)

5. How much space, that might otherwise be income-producing, will the proposed equipment require?

6. What effect will the proposed system have on the cost of the building itself?

7. Is it feasible technically and will it be profitable to automate the system?

8. Has he qualified himself to judge the mechanical and electrical adequacy of various makes of equipment, comparatively?

9. Are architectural considerations preventing him from selecting the system he considers best for the job?

What is needed for the designer to answer the first eight of these questions effectively? In a dynamic industry such as air-conditioning could and should be, it means he must be not just a design engineer but also a *business* engineer. It means actually visiting and examining the latest installations—particularly if they are unique—and interrogating those most intimately concerned with them. How often? Once every two or three years. Where? Certainly in those parts of the country where climatic conditions are comparable to those at the location of the proposed job.

Are designers doing this? With but few exceptions, the answer is *no*. The cost in time and money is one reason. The lack of incentive is another. It is your money, not his, that is being spent. But he only gets a little of it. If the architect's commission is 6%, that may be the commission that the consulting engineer will receive on the mechanical contract. The architect would, however, be less than human if he didn't want to conserve as much of his total commission as he reasonably can. That is why it is not uncommon for consulting engineers to work at a commission of only 4 or 5% of the mechanical contract.

Such engineers face a dilemma. They cannot possibly assume the expense of carefully considering a number of alternative designs for each job. They cannot afford to go into detail sufficient to isolate the one system which will reasonably satisfy the building occupant at the least total cost to the owner, over its probable life expectancy. But even if they could, they might be prevented from using that system because of architectural considerations.

Alas, since the architect cannot even find, let alone engage, the services of an engineer with the ideal qualifications already implied, what are you, the owner, to do?

First, you should recognize that the engineer is system and building—not people—oriented. He writes articles about satisfying minimum owner requirements. In them he mentions noise, appearance, flexibility of space use, etc. He never even mentions the word *comfort*. Or he writes learnedly about the pros and cons of two systems and ends his article with: "As a conclusion it may well be stated that . . . cost comparison between the two systems is the most important factor rather than the operational cost and *occupant comfort*." (Italics added.)

Second, when engaging the architect, the owner should ask to whom he plans to give the engineering. Then, why not interview this engineer, in the architects' presence before he is appointed? Why not ask him the foregoing questions? The owner and the architect may both be helped by this. And so may the engineer.

Third, when you talk with the engineer ask him to kick the mystery bit—wet bulb, apparatus dew point, grains of moisture per pound, etc.—and talk in terms the ordinary person can comprehend. Example: You want to know where the conditioned air will enter the occupied spaces—at the window sills, from the ceiling, from grilles high on the walls? To you, where it enters is clearly the inlet. He calls it the outlet—that's where the air comes out of his system.

Fourth, remember that the engineer wants his competence challenged. He wants constantly to prove to his fellow engineers and especially to himself that he knows his stuff. As one of them said to me, "I design four or five jobs with the XYZ System (one of the best, yet the simplest of all to design). Then I have to prove to myself that I am still an engineer, and let myself go." "At the expense of the poor owner," I added. Yes, engineers, too, fall prey to the "look ma, see what I done!" syndrome. Just don't let him do it at *your* expense.

Consulting engineers insist that their shortest commodity, if not money, is time. Yet, they often oppose the use of simple systems. Such systems often save 50% or more of the man hours needed for the job and thus of the job execution cost. This would be far more than they might lose in lower commissions resulting from the reduced first cost of such simpler systems, thus making such systems actually more profitable. But, perhaps subconsciously, they seem to fear that, if a system is too simple, the architect will design it himself. To prevent this, they sometimes "accomplish by extremely complex roundabout means what actually or seemingly could be done simply," to borrow Webster's definition of a *rube goldberg* contrivance.

When engaging an architect, the owner should find out exactly what services the air-conditioning design engineer is to perform. Of course he will calculate the loads; write specifications and show a layout of the system on the building plans. But, to what extent, if any, is he being paid to participate in these other necessary engineering functions?

1. Contractor Selection—Knowing from experience the relative qualifications of the bidding contractors, will he have a voice in evaluating the bids and awarding the work?

2. Coordinating Problems—Such problems often arise between, for example, the structural and the mechanical engineers. Will these be resolved by the architect, with possibly adverse effects on the functional adequacy of the comfort system, or will the engineer participate in resolving such conflicts?

3. Job Supervision—Is it agreed that the engineer will have a competent man on the job to assure the correct installation of the system or will this be one of many responsibilities of the architect's *job clerk?*

4. System Balancing—Some systems, particularly the older ones, require a great deal of system balancing to assure that the proper amount of air or water reaches each location. Doing it right is not easy. Who will do it: the contractor, the consulting engineer or an outside system balancing firm? And, who pays?

5. Start-up and Operating Instructions—Most quickly to discover and eliminate its initial bugs, the man who designed the system should participate in its start-up. In addition, he should prepare operating instructions and wiring diagrams that will be understandable by the technician, custodian or janitor who will operate the system, unless it is automated. And he should be required personally to instruct this man in its proper operation.

The consulting engineer cannot escape responsibility for a large portion of the complaints that come from people who like comfort but not air-conditioning. Is he thoroughly familiar with all modern systems? Has he directly or, through the architect, clearly conveyed the comfort limitations of his design to the owner? Has he insisted on full responsibility for the entire air conditioning job or has he accepted a cut fee for doing only a part of the necessary work, leaving the rest undone, at the mercy of the contractor or in the inexperienced hands of the architect's representative?

How, then, can the owner be sure to get that system which, if it is in a commercial building, contributes most to the bottom line of his profit and loss statement? How can he get that system which has the lowest total owning and operating cost and that still retains the ability to attract customers, tenants, guests or patients by offering them a higher degree of comfort than is offered by competitors? Several suggestions appear on page 198.

In the engineering department of Howard Hughes' Aircraft Factory hangs a large sign. Its message is as simple and ungrammatical as it is clear. It reads: SIMPLICATE AND LIGHTNESS ADD. The office of each air-conditioning design engineer needs a similar sign, one that reads: SIMPLICATE, COST REDUCE AND COMFORT ADD.

THE PUBLIC UTILITIES

A Chicago architect—unusual because of his technical understanding and appreciation—designed a new high-rise office building for a nearby city. The air-conditioning system he selected was simple, modern, economical and had the capability of providing full-time comfort for all occupants during all hours of occupancy. What happened? The local gas company got into the act. If the comfort system would go *all gas,* they would finance it with 5% *money.* That is the way it went. The compromise system that was installed fell far short of providing full-time comfort during *normal* hours of occupancy, not to mention after hours or on weekends when, because of its design, it was completely shut off.

Obviously, the gas company went beyond its normal responsibility of distributing and selling an excellent fuel. It did exactly what electric utilities do, it went hard after the business. The single fuel utility—the one that sells gas only or elecricity only—is usually more aggressive than those who sell both. Most dual fuel utilities do not get excited unless the third energy source—oil—is being seriously considered. But, there are exceptions. San Francisco's Pacific Gas and Electric Company has, so far, pushed gas over electricity for comfort systems. Upstate New York's Niagara-Mohawk Power Corporation, in contrast, pushes electricity even in those areas where it also sells gas. (Discussions of oil vs. gas vs. electricity appear in Chapters 12 and 13.)

There is no harm in working one utility against another to get the best deal from each. Even if the gas company's deal appears best, it would be a mistake not seriously to *consider* going all-electric, particularly in the case of multiroom structures. Warning: Make sure that the engineer designing such an alternative system understands and selects a system that capitalizes fully on the things that electricity, but not combustible fuels, can do.

THE MANUFACTURER

Air-conditioning systems are not manufactured, they are assembled. "Large systems maybe," you say, "but not my little blow hard." Yes, your room cooler too, was assembled. Chances are the thermostat was made by Ranco, the fan motor by Redmond, the fans by Torrin, the control switch by Arrow-Hart and the heart of the system, the hermetically sealed compressor, by Tecumseh, but with a GE motor. What, then, was done by the factory from whence it came?

One chef can take an assortment of food stuffs and create a meal fit for a king. Another, using the same assortment, ends up with an inedible mess. Did the maker of your unit know his stuff? Did he pick the finest components? Did he pick them plenty large and make it *hell for stout* by mounting them on 16-gauge zinc-coated sheet metal as against 20-gauge black steel? And was he careful with his assembly work, processing and testing? If so, you should have good results, provided the dealer sold you the right size. But if minimum factory cost was the maker's maximum objective, then you have a noisy *window shaker* that may not last much longer than the 5-year warranty that usually covers the compressor.

So, too, it is with most large systems, except that the design engineer, instead of a factory engineering department, sizes and picks the components. The contractor then puts them in place. In the first case, the manufacturer whose name appears on the unit is responsible for its performance. In the second, responsibility is divided. If the results are poor, "who done it"? Is it the fault of the consulting engineer's design? Was it properly installed? Is it being operated as intended by the designer? Are the major components delivering their rated capacities?

Trying to fix the blame for improper results from a field-assembled system can be frustrating. One repetitive gambit is a version of the eternal triangle: The unhappy owner goes to the major supplier—the one who made the heating, cooling and air handling equipment—and, of course, he pleads innocent. The trouble, he explains patiently, is with the controls. The controls were sold by another manufacturer, on a separate contract, and on an installed basis. The harrassed owner goes to him. What does he get? A clear, logical, convincing explanation that, "it's not the controls, it's the air-conditioning equipment". In the meantime, the building occupants, all of whom like comfort, complain bitterly about "that damn air-conditioning".

Progress, contrary to conventional wisdom, is not the product of the largest producer in a given field. It wasn't General Motors that introduced 4-wheel brakes. It wasn't GM that led the switch from wood frame to all-steel bodies. Nor were they first with safety glass, seat belts or even aluminum pistons. For the fields they cover, the same may be said, collectively, of the *big three* of air-conditioning. Having, in general, done a good job doesn't relieve them from criticism in five areas:

1. They have been slow to put their weight behind satisfying the desires of an increasingly sophisticated public which has long demanded better comfort conditions.

2. They have put great effort on improving their products and/or reducing their cost. But too much of it has been spent on the components of inadequate systems which, functionally, have not been improved significantly since first introduced 25 to 40 years ago.

3. They have been reluctant to develop, produce and promote the sale of new systems that provide more comfort for less money, for fear of making obsolete what they now have.

4. They have let intense competition lower their standards. A lower price is obvious to the buyer; higher quality often is not. Lower price calls for lower cost. Value analysis is used to achieve it. This is the attempt to reduce manufacturing cost through design changes *without detracting from product quality.* It leads to *design brinkmanship,* with much of the true value often analyzed out. Related to this is the built-in obsolescense that rewards the manufacturer with premature replacement sales at the expense of the unhappy owner.

5. They have failed fully to capitalize on the opportunities offered by the greatest single development that has been made since the first successful mechanical refrigeration plant was put into operation—the true hermetically sealed refrigeration circuit (page 93).

There are several reasons for this, some good and some not so good, but all understandable.

The air-conditioning industry has been growing so rapidly that it is natural for managements to ask, "why change a winning game?" To change the game is to make large present investments in engineering designs and factory facilities obsolete. Worse yet, after years of training and experience, the sales force, the consultants, and the contractors have learned how to design, estimate, apply, install and service systems which have long been available. Who wants to tell them to forget what they have learned and to start over again?

It is not easy for a large, long-established pioneer suddenly to change his pitch. It is not easy for him to escape the well-known hazards of long established competence. It is even harder for him to adopt what was first developed and introduced by some small, unknown upstart. To a degree, all product design departments and even top managements suffer from NIH syndrome—Not Invented Here. It is not easy to reject one's own propaganda, even after it has become obsolete.

This, of course, is what gives the alert, eager, wide-awake, industrious young upstart his opportunity to get into the game. But the owner and his advisers usually feel more comfortable in dealing with the large, old firm with the big name. The man who finally has the monkey on his back is playing it safe when he does this—when he goes for what is already installed by such a firm in the fine new building down the street. And thus the future tenants are deprived of the comfort they have, in this day and age, every right to expect.

THE MONEY MEN

The First National Bank of a large midwestern city will be the principal tenant of a fine, new, 40-story office tower. As such, they have a strong voice in its design. Contrary to usual practice, the senior officer in charge of the project personally made a thorough investigation of comfort systems. He checked on the results being obtained from traditional systems. But he also checked on latest developments. Then he made his decision. What happened?

He convinced everyone concerned, including the developer, that the new system he had selected was the right answer; the right answer provided the unique, demonstrably superior product of a specific manufacturer was used. He convinced everyone, that is, except the insurance company that was putting up the mortgage money. A few years before, that insurance company had financed a beautiful new high-rise structure on Chicago's lakefront. The same system, allegedly, had been used there. Its equipment was the product of a company with a better known name than that of Spiro Agnew. The equipment was also some of the flimsiest on the market. Its sole virtue had been low first cost. It came complete with high noise level, limited life expectancy and built-in obsolescense at no extra cost. And it soon proved to be unsatisfactory.

What did the insurance company do? It arbitrarily vetoed the use of the new system in the First National Bank Building. And this without regard to the demonstrably superior equipment that would have been used—quiet, reliable equipment with an exceptionally long life expectancy. So the building occupants will have to make do with the part-time comfort that will be supplied by the more costly traditional system that has been selected. This, then, is simply one example of how, unwittingly, money men may deprive those who, in the end, will pay off the mortgage, of the full-time comfort for which they are paying.

WHAT ABOUT ASHRAE?

The speaker was addressing a chapter meeting of the American Society of Heating, Refrigerating and Air-Conditioning Engineers. His audience consisted of some eighty architects, consulting engineers, contractors, owners, manufacturer's representatives and others interested in air-conditioning. Looking his audience in the eye, he asked a simple question: "Who in your opinion are the three most important individuals concerned with any large air-conditioning project?" Gradually he was able to cajole answers from the audience. The consensus was the architect, the consulting engineer and the contractor.

Feigning outrage, the speaker shouted, "You are wrong! First comes the occupant of the conditioned space, then the operator of the system—who gets the complaints from those who do not like its results—and with him its owner."

Even ASHRAE is building, not people, oriented. This is confirmed by the official definition of air-conditioning adopted by this Society:

> Air-conditioning: The process of treating air so as to control simultaneously its temperature, humidity, cleanliness and distribution to meet the requirements of the *conditioned space.*

I have italicized the two words with which I disagree. I submit that the conditioned space doesn't give a damn. To satisfy my concept of air-conditioning the last phrase should be "to meet the requirements *of the occupants* of the conditioned space".

An official of an experienced firm of real estate developers was put in charge of the construction of a new downtown shopping plaza and high-rise office tower. His company had been persuaded to try out a unique new comfort system in two small office buildings they had constructed elsewhere. The comfort results, despite a few problems, had been excellent. A leading firm of New York City consulting engineers was engaged. They were reluctant to specify the new system. It was not one of the office building systems covered in the contemporary edition of the ASHRAE Guide.

Later, this developer publically stated—and in subdued form it appeared in a trade journal—that he felt ASHRAE had let the consulting engineers down by not providing sufficient data to enable them to engineer and specify such a new system with confidence. The point he missed is that the ASHRAE Guide leads its industry in the field of fundamental data required by the design engineer in estimating loads. But, when it comes to system design, it follows the art. In my opinion, this is the only position it can take if it wants to hold together, in one Society, the vested interests of the industry with the venturesome entrepreneurs with whom so many of the progressive, new ideas originate.

By the time the industry generally adopts and the ASHRAE Guide is free to reflect ideas that *now* make good sense, there will undoubtedly be an accumulation of newer ideas that make still more sense. There is, however, a gleam of hope on the horizon. The industry, as reflected by ASHRAE, has recently been engaging in useful self-criticism. An example of this are these excerpts from a talk given by past president Daniel D. Wile, plenary speaker at ASHRAE's 1969 annual convention:

> It's time we let the public know that we know how to provide good all-year air-conditioning. . . We must admit, too, that even today we are doing a particularly poor job on small air-conditioning applications—residences, motels, hotels, small commercial and apartment buildings—because they cannot cope with the in-between seasons. . . Today's apartment projects—with excessive noise, drafts, poor spring and fall control and no provision for outside air.

He further indicated that this would make it appear that the industry is moving backward instead of forward.

He then proposed that ASHRAE define the various types and classes of systems in terms of what could be accomplished in the way of comfort throughout the year. He particularly urged an evaluation of their ability to maintain comfort during mild and humid weather. His views are increasingly being shared by others. Thus, at its convention of January, 1971, ASHRAE, by means of a symposium, took its first public steps toward developing a method for rating air-conditioning systems in terms of their comfort capabilities and economics. One of the nice things Dan Wile could have said about the comfort industry, but didn't, is the ease with which so much of it can be improved upon.

OTHER VESTED INTERESTS

The comfort segment of the air-conditioning industry is relatively new. It is only half as old as the automobile industry. When an industry is new it is innovative. As it grows older, it begets vested interests. They cling to the status quo more tightly than do barnacles to a ship. Together, these interests endeavor to make eternal what no longer makes sense. As we have seen, the air-conditioning industry is no exception.

Take architects and consulting engineers. As already indicated, they usually work on a commission. Let us say it is 6% of the job cost. They can not completely ignore the fact that, if the comfort systems costs $3.33 per square foot, their share is 20¢. If it costs $10 per square foot, it is three times that amount. What incentive have they to spend less than the maximum the owner will go for—or a little more? After all, they are quasi-purchasing agents who, in effect, are paid a percentage on all they can persuade the owner to spend.

As already indicated, the long established manufacturers who created the air-conditioning industry have a powerful vested interest in maintaining the status quo.

What about the contractors who make the installations?

"N. J. Contractors Issue 'Battle Kit' To Fight For Hydronic Heat,"[3] reads the headline of one trade paper article. It describes the kit favoring hot water heating that the Mechanical Contractors' Industry Council of New Jersey distributed to mechanical contractors and local unions throughout their state.

"St. Paul Piping Group Fighting Electric Heat on School Jobs,"[3] reads another headline. What would you do if, after having built a pipe-fitting business, your tax dollars were being spent for heating/cooling systems that required no pipes? Or, having built a sheet metal business, air ducts were no longer required? When that is about to happen, the contractors and building unions bury the hatchet. Contractor groups and labor unions have resorted to boycotts and law suits to prevent the installation of systems that do not promise them as much work as they have traditionally done.

There is one thing in the comfort business that has been conclusively proved: when using components of comparable quality, the more of the work that is done in the factory and the less that is done in the field, the better the quality of the finished job and the lower its cost. But even the U. S. Supreme Court has helped the leadership of the building trades unions to abuse their economical and political power and block technological progress. It has ruled, in effect, that these workers are entitled to retain field work they have traditionally done.

Maximum comfort at minimum cost? Who cares about that? Well, you should. It's better for you than minimum comfort at maximum cost. Maurice Maeterlinck spoke the truth when, in The Bluebird, he said (using my imprecise translation from the French), "At every door that leads to human progress there stand 1000 men, self-appointed to preserve the past."

THE SERVICE CRISIS

"The hot summer has produced a major air-conditioning crisis in the Washington area. One dealer recorded 1,000 calls for service in one day." Some apartment dwellers, "have been without air-conditioning more often than with it all summer. People with window air-conditioners have to wait from two weeks to two months for service."

That is how the Washington Evening Star said it on August 11, 1968. But it was not only Washington. Nor did the problem begin or end with 1968. Nevertheless, the Washington situation has been so bad that a bill was introduced in Congress by Representative Charles S. Gubser to empower the District of Columbia to, "make regulations to prescribe standards relating to capacity, operation, and maintenance of equipment used for air-conditioning apartments."

Let us examine the major reasons for this appalling condition:

1. Some dealers and contractors consider service work unprofitable.

2. Shortages of parts and components, due to low inventories all the way from factories to service departments, aggravated by strikes at manufacturing plants.

3. An insufficient number of service mechanics to meet the demand—the major reason.

4. Repeated *call backs* on equipment which was previously passed as being repaired, because of limited knowledge and skill of the service man.

5. Union practices restricting the geographical areas where apprentices may work.

That is one side of the coin. The other side is the owner who neglects his equipment or doesn't realize it should be regularly checked over, until it breaks down. That is most apt to happen during a heat wave, when it is needed most. Some manufacturers, dealers, contractors, and service companies offer annual (or longer) contracts covering both preventive maintenance and emergency service, both labor and parts. Owners who have such contracts are taken care of first, when trouble occurs. The others must wait their turn.

The cooling elements of a comfort system, being more complex, are much more apt to need service than the heating portion. That is why some dealers lay off their service men at the end of the cooling season. The best men easily obtain other work and are lost to the industry. An example will show one way to mitigate the service crisis:

My phone call was to the head of a leading air-conditioning contracting and service company in New York City. "Jack," I said, "do you realize that the job you installed in the 72 Wall

Street Building is now three years old." He did and I continued, "As you know, we always recommend that all the equipment be put through our 25-step preventive maintenance program every three years. We now want to persuade the owner to have it done. What will you charge to do it?"

What was his reply? "When would you want it done?" Then he explained: "In order to hold my best men, I have to pay them all winter when they have nothing to do. The others I let go at the end of the rush season." The upshot of our conversation was this: if they could do the job in January and February the price would be so much—and he mentioned a remarkably low figure. As the season advanced, the price would rise and at no price would they do the job in July. Then they would have so many emergency calls they couldn't possibly do anything else.

The people in Washington—and elsewhere—who can't get prompt service when their cooling stops during a heat wave still like comfort but, at the time, they despise air-conditioning.

Who, then, is responsible for the air-conditioning discomfort which causes so many complaints? It is easier to ask "who isn't"?

Even the occupant may be at fault. For instance, a typical guest at a well-known chain motel.

This guest, sticky and weary, had arrived late on a hot afternoon. The first thing he did, on entering his hot room, was to push the ON button on his incremental all-electric conditioner. Then he turned the thermostat knob as far as it would go in the direction of COOLER. By the time he had showered and changed, the room felt good. Life was worth living again. With a smile of anticipation, he headed for the cocktail lounge to relax.

Some three hours later, having dined, he returned to his room. He would get out of his clothes, watch Gunsmoke and The FBI on his TV, then turn in for a good night's rest. But, by then, his room was too cold. So, with a grumble, he went over to his air-conditioner and gave the thermostat knob a good twist in the opposite direction, toward WARMER. Soon he was so engrossed in his program that room temperature was out of his mind. Then, with a start, he said to himself, "God it's hot in here." That was when he gave the thermostat knob another twist, in the opposite direction, and called the front desk. The control system was so simple it had confused him. He was doing what so many people do—over-controlling.

Chapter 9

Buying Carryout Comfort

High above Fifth Avenue, there it hangs, unsightly evidence of man's determination to defeat discomfort. A sword of Damocles, it is perched precariously above the unsuspecting heads of pedestrians hurrying along the sidewalk below. The screws with which it is fastened are rusting away. How long will they hold it in place? Who knows! And if it lets go, it won't be the first one to do so.

It is not alone. It has thousands of companions of assorted shapes, sizes, ages and makes. Such architectural merit as was once possessed by the buildings they infest, they have long since destroyed. In summer, rusty water oozes down the spandrels beneath the dirty windows, the cleaning of which their presence has forestalled. And so the buildings, with their stained facades, become prematurely old. Yet, on that rare, hot day when the sun is bright and the sky is blue these eyesores do make themselves known to the people below. Every so often, someone looks up and says to his companion—or to himself—"I thought it was raining," as a drop of water impinges on his bare head.

Our bodies were moist from walking in the heat. We took the elevator to call on a friend. As we were ushered into his private domain, we get another view. But not of the outdoors. Centered on the window sill is a cabinet. It has a face of juke box art, embellished with a gang of multicolored push buttons and two gold-like knobs. Two dirty beige wings fill the space between it and the window frame. It is making so much noise that we all speak louder, to be heard. My friend smiles an apology. "I've got it on low," he says. We visitors are both polite. "Yeah," one admits, "it's nice and cool in here." "Yeah," says the other, "your window unit is doing a good job." It, too, was a long-time coming.

THE ROOM-COOLER IS BORN

In 1931, at Monterey, I spoke before the California Association of Ice Industries. My mission? Motivate them to sell room coolers. Icemen sell room coolers? Yes, indeed, Carrier's first.

Mounted on casters was a mahogany grained metal box maybe 45″ long by 24″ wide by 32″ high. The upper half of one end was a door that doubled as an air inlet grille. At the other end, a motor-driven fan was arranged to blow upward through discharge grilles. The cooling source was noiseless and troublefree —a 300 lb. block of ice. It rested on the edges of lengthwise aluminum fins, designed for uniform cooling as the size of the block diminished. Under the fins was a tank to catch the meltage, with a connection for draining.

It didn't put Carrier in the black. Only a few hundred were sold—mostly in hospitals. But it worked. Charged with ice in the basement, it would be taken up in the elevator and wheeled to a room. There, on a sweltering day in July, it offered benign relief to the fevered patient.

Broad coverage on air-cooled, mechanically refrigerated room coolers was granted to two Philadelphia inventors in 1932.

The Depression soon killed off the few small manufacturers who tried to produce them. But soon, several that kept going were at it: York starting in 1934, Carrier in 1936. It didn't make them rich. From a total of perhaps 6,000 in 1935, production grew to 13,350 in 1938, 31,339 in 1941. Then it was stopped *for the duration.*

Early designs were *consoles*—upright and floor-mounted— typically 36″ wide, 18″ deep, 40″ high. A short, divided duct was fastened at the rear, just high enough to clear the window sill. Outside air was sucked in through one side, pushed through the condenser, then discharged 20° warmer to the outdoors, through the other side, carrying the room heat with it. Often the duct was arranged so that the window could be closed when the unit was not in use or opened wide for washing. Often a radiator had to be removed so one could be installed. Thus a few were also offered with electric heat. However sold, they were expensive.

Just before Pearl Harbor, the window unit was born. It featured low cost, easy installation, no radiator removal, condenser outdoors, no duct needed, and, to hell with the outdoor looks. By 1946, after hibernating five years, consoles and window units were again in production—43,000 in 1947, 194,000 by 1950. Until then, the design of the units had been controlled by engineers. Production was largely in the hands of engineering-oriented manufacturers of air-conditioning equipment. Then, irresistibly, Vance Packard's Waste Makers took over. The console became extinct and the window unit became a household appliance.

Everybody got into the act. By 1954, there were 133 brands on the market, turned out by perhaps 35 factories. Aided by quirks in the Federal Excise Tax regulations, capacity ratings lost their meaning. Each season, the last firm to issue ratings had the highest ones. Unbelievable wheeling and dealing was used to market window units. Factories agreed with dealers to *buy back* units unsold at summer's end. If early summer was hot, there was a shortage; if chilly, a huge carryover. And, much of the time, price cutting was rampant. Soon, it seemed to me that the window unit was like the beautiful girl that became a prostitute before she reached the age of consent. And the two had something in common. They both provided temporary relief.

Temporary relief? Well, sort of. In 1951, a typical window unit rated at 9000 Btuh weighed over 200 pounds. It was built to last. A few years later, one rated the same and carrying a name that every house needs, was down to 93 pounds. This cut life expectancy about in half—seven or eight years instead of 15 or 20. As I lifted one, the whole thing bent enough to make a frightful noise as distortion let the fans hit the nearby metal. But it worked. And the cost was low. Production soared. In 1953 room coolers turned the corner. That year alone, sales matched the entire previous postwar output—over one million units.

That was the year the room cooler was discovered by the American Public. That year Fortune ran a piece on The Air-Conditioning Boom. In Reader's Digest, J. P. McEvoy told how window units "blossoming in the windows of apartment buildings and stately palaces" had put bustle in pre-Castro Cuba, and how a senora gurgled about the one in her Havana bedroom: "Pepe was never so peppy!"[17] Millions, said the New York Daily News, are joining the non-sweat set. From one million window units in 1953, factory shipments broke through the four million mark in 1967. In 1970, they set a new record—over 5.5 million units for a retail value of about $1 billion.

WHAT MADE IT ROLL?

It was at a convention in Atlantic City at the end of 1949. Standing together, three of us were shooting the breeze. One was an officer of ASHRAE. To him I remarked, seriously, "At our next convention, someone should present a paper titled: *The Case For the Single Room Air-Conditioner.*" Before he could reply, the third man, the brilliant chief development engineer of a great air-conditioning company, drew himself up to his full six-feet-one and sardonically asked, "Is there a case for the single room air-conditioner?" Indeed there is, even if engineering perfectionists don't like it. That is what made it roll. This case can easily be condensed into five major points. Each of them is so obvious that it would be superfluous to belabor them farther. The window air-conditioner:

1. Provides the least costly retreat from summer's heat that money can buy—both first cost and operating cost.

2. Can be installed more quickly and easily than anything else that will provide the same summer benefits.

3. Is operated under the complete control of the individual occupant; he can have cooling anytime he wants it—even during the February thaw.

4. Is ideal for the renter. He can take it with him when he moves.

5. Because it is a removable *appliance* it may easily be bought on monthly payments.

Point No. 3 reminds me of a visit to the head office of the Glidden Company, in Cleveland, some years ago. We wanted to air-condition their entire building. It was early April. A chilly breeze was blowing from the lake. The official we were calling on took us into his private office. Under its single window was a radiator. It was hot. On the sill above it was a room cooler. It was blowing its heart out, in a futile effort to compete with the sizzling radiator.

After we were well enough acquainted for me to do it, I raised the question as delicately as I could—something like: "Why the hell do you have the radiator on for heat and the window unit for cooling, both at the same time?" His answer was more sensible than economical, "With the radiator on I get too hot. With it off, I'm cold. The room cooler, operating on its thermostat, gives me just the temperature I want!"

WHY NOT FOR HEATING, TOO?

Ever since the early days, console and, later, window models were offered with electric heat—available optionally *at slight extra cost*. Enough heat to carry you through the winter in Minneapolis? No, but more than enough to do it in Miami. And enough to do it before the central heating plant is turned on in fall, after it is turned off in spring.

Now, a few manufacturers offer window units (for window or through-wall installation) with increased heating capacity for winter use in all parts of the country. Where they use electric resistance elements and are installed under the window, i.e., through the wall, they do a good heating job. Their noise level, however, is not always acceptable in winter. Some motel chains use them, particularly those that put top emphasis on low first cost.

Some years ago, several makes of room coolers with *reverse cycle* (or heat pump, it's the same thing.) heating were placed on the market. Their performance was not satisfactory. However, some manufacturers may still be offering them, though possibly only for use in mild climates. My recommendation is this: do not get involved with air-cooled reverse cycle room coolers.

SO WHAT SIZE DO I NEED?

Go to *Appliances* in your favorite department store. Ask the salesman what size you need. Chances are good he'll put his hand on a sample and say, "This one will cool a room with a floor area *up to* 300 square feet," and be telling the truth. But he may not say—and may not know—that the unit is too small for some rooms of half that size. It all depends on your room, your climate and how it will be used. That determines how many Btuh must be removed on a hot day to keep you cool.

Certified room coolers (about which more later) are rated on their data plates in Btuh. This is given for only one set of out-door and indoor conditions: 95° and 45% rh outdoors; 80° and 50% rh indoors. Capacity will be less when it is hotter outdoors; more when it is cooler. But, if you want it cooler indoors, you will lose some capacity—maybe 5% less at 75° than at 80°.

Should the cooling power of the unit match the load—meaning the number of Btuh that must be pulled out of the room to keep it cool? Yes and no! Are you going to leave it run continuously? Are you willing to set the thermostat 3° to 5° cooler than you really want it? If so, pick one 20% smaller than the load. Flywheel effect (page 18) will carry you over the afternoon *heat peak*. And this will give you the best possible dehumidification. But, if you don't want to turn it on until you get to the

office or come home from work, even in the hottest weather, get
the one with a capacity no less than the load and maybe a little
larger.

But how much is the load? Let's figure it for two offices that
are alike. Neither is on a building corner. Both are the same
size, 12.5 feet wide by 16 feet long—that's 200 square feet. The
ceiling height is 8 ft. One office faces west, the other north. The
windows in each are ten feet wide by five feet high or 50 square
feet of single glass, with drawn venetian blinds. The offices on
both sides are air-conditioned. So is the corridor and the offices
above and below.

So here, in Table 9-1, you have it. Since all surrounding space
is already cooled, a 4000 Btuh unit is more than enough for
north. For *west* you need over twice as much. A 10,000 Btuh
unit would probably be the nearest commercial size. (The
smallest unit listed in the 1971 AHAM directory is rated at
4600 Btuh. More about that, later.)

ITEM	QUANTITY	BTUH Each	TOTAL BTUH West	North
West Window (facing sun)	50 sq. ft.	79	3950	——
North Window (in shade)	50 sq. ft.	14	——	700
Outside Wall	12-½ lin. ft.	30	375	375
People, seated (includes Vent. Air)	2	600	1200	1200
Lights	400 watts	3.4	1360	1360
TOTALS PER WALL			9885	3635

Table 9-1—Simple Cooling Load Estimate

From this, it is obvious that because two rooms are the same size is absolutely no assurance they will require room coolers of the same cooling power. Tables 9-2 and 9-3 have been computed using AHAM's Cooling Load Estimate Form for Room Air-Conditioners, with room exposures as shown in Figure 9-1. These tables give accurate answers, *provided* the

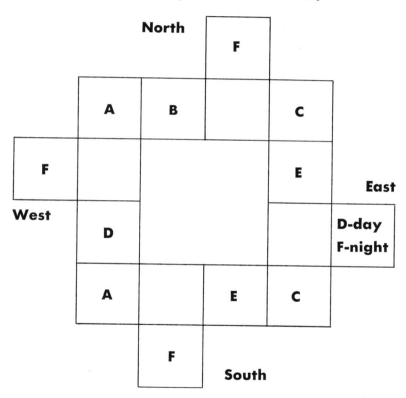

Figure 9-1—Room Plan Diagram

assumptions that follow fit your case and provided further that
you observe the suggestions illustrated in Examples 1 and 2.

ASSUMPTIONS

1. Summer temperatures do not often exceed 95° or you

NORMAL CEILING		FLOOR AREA sq. ft.	HOT CEILING	
25% Glass	50% Glass		25% Glass	50% Glass
ROOM A				
5290	7440	120	6970	9120
6190	8620	170	8570	11,000
7400	10,300	240	10,800	13,600
8820	12,300	330	13,400	16,900
10,600	14,600	460	17,000	21,000
ROOM B				
3550	3860	120	5230	5540
4140	4510	170	6520	6890
4950	5380	240	8310	8740
5900	6420	330	10,500	11,000
7200	7800	460	13,600	14,200
ROOM C				
4740	6340	120	6420	8020
5540	7320	170	7920	9700
6620	8720	240	9980	12,100
7900	10,400	330	12,500	15,000
9520	12,400	460	16,000	18,900

Table 9-2—Room Cooling Loads in Btuh, Daytime Only

multiply your Btuh selection by the factors shown on the AHAM map, Figure 9-2.

2. The building is of reasonably good quality frame or masonry construction.

3. NORMAL ceilings are under occupied space, under an

NORMAL CEILING		FLOOR AREA	HOT CEILING	
25% Glass	50% Glass	sq. ft.	25% Glass	50% Glass
		ROOM D		
5050	6860	120	6730	8540
5900	8050	170	8280	10,430
7050	9580	240	10,400	12,900
8420	11,500	330	13,000	16,100
10,100	13,600	460	16,600	20,100
		ROOM E		
4430	5620	120	6110	7300
5180	6590	170	7560	8970
6190	7860	240	9550	11,200
7380	9380	330	12,000	14,000
8920	11,200	460	15,400	17,700
		ROOM F		
5600	7960	120	7280	9640
6550	9350	170	8930	11,700
7830	11,100	240	11,200	14,500
9340	13,300	330	14,000	17,900
11,200	15,800	460	17,600	22,200

Table 9-2 (Continued)

insulated roof or have at least two inches of good ceiling insulation.

4. HOT ceilings are not insulated and are adjacent to an uninsulated roof.

5. Ceiling heights are not over ten feet.

6. The width of the room is not less than 2/3 of its length.

NORMAL CEILING		FLOOR AREA	HOT CEILING	
25% Glass	50% Glass	sq. ft.	25% Glass	50% Glass
ROOMS A and C				
4270	4990	120	5950	6670
5080	5820	170	7460	8200
6200	7060	240	9560	10,400
7540	8580	330	12,200	13,200
9360	10,600	460	15,800	17,000
ROOMS B, D and E				
3960	4270	120	5640	5950
4720	5090	170	7100	7470
5770	6200	240	9130	9560
7020	7540	330	11,600	12,200
8760	9360	460	15,200	15,800
ROOM F				
4580	5510	120	6260	7190
5440	6550	170	7820	8930
6630	7920	240	9990	11,300
8060	9620	330	12,700	14,200
9960	11,800	460	16,400	18,200

Table 9-3—Room Cooling Loads in Btuh, Nighttime Only

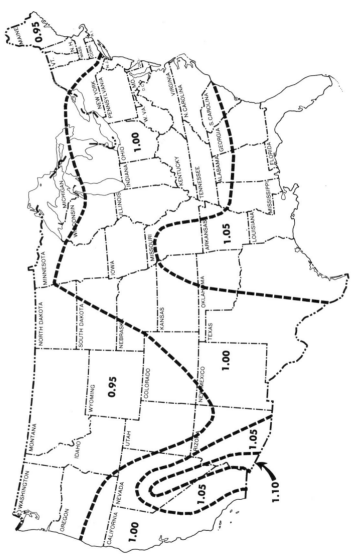

Figure 9-2—USA Room Cooler Capacity Factors

7. Doors and windows are closed. (Add 300 Btuh per foot for doors or arches continuously open but not over 5 ft. wide. If wider than this, consider the two rooms as one space.)

8. Lights are off when the sun is shining, on when it isn't. (Table 9-3 assumes lights are on and amount to one watt per square foot.)

9. Direct sunshine is excluded by venetian blinds, light colored drapes or outside awnings.

10. All outside walls have windows.

11. The number of people in the room is normally two. (Add 600 Btuh for every additional person at rest. This includes minimum extra ventilation.)

12. Adjoining spaces are not cooled.

Of course, the assumptions won't all fit your case, so use good judgment. If you keep lights on when the sun is shining, add the wattage of the bulbs and multiply it by 3.4 to get Btuh. (Figure 50 watts for each 4-foot, 40-watt flourescent tube. That covers the ballast, too.) West (Table 9-1) has its 12.5 foot wall facing west. If the 16-foot wall faced west you would need considerably more cooling. If the unit is for a hospital room, check for exposed steam pipes. They are apt to be hot in summer —to serve sterilizers.

EXAMPLE 1

This is an office in the same building as *west* and *north*. But its size is 16 x 16 ft.—256 square feet. And it is on the southwest corner of the top floor. Neither its ceiling nor roof are insulated. The adjoining spaces are *not* cooled. Since the office is for DAYTIME use, pick the room cooling load from Table 9-2, following these steps:

1. Locate the office on Room Plan Diagram, Figure 9-1. It is an *A*.

2. Under FLOOR AREA, Room *A*, find 256 square feet. It doesn't show, but 240 does. For the moment, let's use that.

3. Since there is no roof or ceiling insulation you will find the correct capacity in one of the columns headed HOT Ceiling.

4. The windows on both outside walls are five feet high and twelve feet wide, thus have an area of 60 square feet. Each outside wall is 16 feet long and 8 feet high, thus has an area of 128 square feet. Hence, the windows have almost half as much area as the walls. Therefore, get your answer from the column headed 50% GLASS.

5. Opposite 240 square feet, you will note the cooling load is 13,600 Btuh. But this office has 256 square feet. That makes it almost 7% larger. Add 7% to 13,600 and you come up with about 14,500 Btuh. A unit in the range of 14,000 to 15,000 Btuh would be a good selection.

DISCUSSION

1. The most heat enters a room through the glass facing west. This gets the hot afternoon sunshine. Therefore, always figure the percentage of glass for the walls exposed to direct sunlight: east, south and west. If it is less than one-third of the total wall area use the 25% GLASS column; if over one-third, use the 50% GLASS column.

What if it is 75% glass? Deduct the 25% GLASS figure from the 50% GLASS figure. For the 250 square-foot room in this case the difference is 2800 Btuh. Add 7% and you come up with 3000. Add this to 14,500 and you come up with 17,500 as the desired capacity of the unit for this extreme condition.

2. Now let us insulate the ceiling and replace half the west glass with masonry or a panel of 2″ insulation. Now what unit does it take? Under NORMAL ceiling find 25% GLASS. Opposite 240 square feet you see 7400 Btuh. Add 7% and you get close to 8000 Btuh. A unit of that capacity would be a good selection. Note how insulation and less glass have cut the size of the unit and thus its operating cost by 45%. But it has also produced a more comfortable office: less radiant effect; less noise because the unit is smaller; less likelihood of objectionable drafts.

3. As stated under Assumptions, Tables 9-2 and 9-3 contemplate rooms which have a length not over 50% greater than their widths. If the room is longer than this and the long wall faces east or west, increase the capacities shown in the tables; if the short wall faces east or west, some reduction may safely be made.

4. If this office were located on a lower floor of the building, such that the west wall is fully shaded after 2:00 PM by a highrise building across the street, the office could then be safely classified as Room C. In that case a unit of 12,900 Btuh would do the job.

EXAMPLE 2

This is a 12 x 14 foot (168 sq. ft.) corner bedroom on the second floor of an old 2-story frame house. The owner has put three or four inches of loose rock wool insulation between the joists of the attic floor. The walls, however, are not insulated. He works hard in the daytime and wants a good night's rest. What size unit does he need?

Step 1—The house faces northeast. His room is on the east corner. The Room Plan Diagram doesn't show this. But it does show Room C on the northeast corner and also Room C on the southeast corner. Since the orientation of this room is between those two, you may safely classify it, also, as a C.

Step 2—Now refer to Table 9-3, NIGHT USE ONLY. In the group headed Rooms *A* and *C,* go down to 170 square feet.

Step 3.—Since the ceiling is insulated look under NORMAL CEILING.

Step 4—Since there are only two small windows—one in each wall—look under 25% GLASS. The answer? A unit with a cooling power of 5080 Btuh. Most manufacturers have units rated 5000 Btuh. One of these should do the job. But, with the next larger size, the room will cool down more rapidly in the evening.

DISCUSSION

1. At 5:30 or so, when the owner gets home from work, that is the time to turn on the unit. By bedtime, not only will the room be cool, but also the mattress.

2. A common practice, for quiet rest, is to turn off the unit and open the window when you crawl into bed. In most parts of the country this makes sense. Why? Because night temperatures usually drop at least 20° from the daytime peak—possibly only 15° at the seashore, but 30° or more in the mountains.

3. If this owner works on the night shift and sleeps in the daytime, he needs a larger unit. From Table 9-2 one of 6190 Btuh is called for.

WILL IT BLOW THE FUSE?

Can the owner plug his unit into the wall receptacle or does he need special wiring? If the data plate on the unit shows an input of 7.5 amps or less (for 115 volts), he can plug it in, according to the National Electric Code. But, if the fuse blows or the circuit breaker pops out, the circuit is overloaded and something else must be turned off. The capacity range of 115-volt units that will satisfy the 7.5 amp limitation is from 5000 to 8000 Btuh.

If a larger unit is needed, it must have a separate circuit from the entrance panel or it will be in violation of the National Electric Code (NEC). Of course, an oversized fuse may be used —at the risk of fire. The largest 115-volt units made by the industry will draw 12 amps and have rated capacities from 7000 up to 14,000 Btuh. Why the 12-amp limit? Because the usual 115-volt circuit is rated at 15 amps. The maximum motor-driven load permitted on a circuit by the NEC is, however, limited to 80% of this or 12 amps.

Why such a broad range in the capacity of units that have the same current input? Because of their design. Small evaporators, small condensers, cheap motors, result in an inefficient unit. But this reduces first cost. The use of large fin coils, more efficient motors, large fans that will handle more air without more noise, will result in a less compact unit but one much more efficient. You will, of course, pay more for it.

WARNING—If you put a 12-amp unit on a separate 115-volt circuit nothing else can be put on that circuit.

Today, most private homes have both 115 and 230-volt service. Some office buildings have the same. Many others have a 115/208-volt supply. Thus many air-conditioners are rated as being suitable for both 208 and 230 volts. But at 208 volts the rated capacity is usually about 3% less than at 230 volts because motors run a little slower.

With power rationing now in vogue, you are better off with a single-voltage unit. It is more apt to keep on running on a hot day, with reduced voltage, than a dual-voltage model. Going abroad to live? Going to take a room cooler with you? Better check the overseas power supply first. About 65% of the electric services abroad are 230 volts, *fifty cycles*. If that is what they have at your relocation, that is what should show on the data plate of your unit. Most manufacturers do produce 50-cycle units. If you put a 60-cycle model on a 50-cycle supply you will have trouble, unless you take a transformer along—one that will cut the voltage from 230 to 200 volts. Exception: Hong Kong has a 50-cycle, 200-volt supply. Our 60-cycle units function well there. But, since all motors run one-sixth slower on 50-cycle than on 60-cycle current, a 12,000 Btuh 60-cycle unit will have a cooling power of only 10,000 Btuh on a 50-cycle supply.

WHAT MAKE SHALL I BUY?

First, buy one that is certified by AHAM. Their 1971 Directory of Certified Room Air-Conditioners lists 54 brands—actually produced by probably 17 manufacturers. The larger producers offer units with a capacity range of from 5000 to 33,000 Btuh.

What does certification cover? Cooling power and electric input: showing volts, amps and watts for each model. Ratings are at an outdoor temperature of 95°, 40% rh; with 80° and 50% rh indoors. Note, however, that your unit may be *rated* at 10,000 Btuh and have an actual capacity of only 9200. Why? Because "cooling capacity . . . shall not be less than 92% of name-plate rating." This is to recognize that not all production units have exactly the same capacity as that of the laboratory prototype. Tables 9-2 and 9-3 take this into consideration.

Heating capacities, if any, and noise level are not certified. If you are worried about the noise level, try to listen to the same model in a location similar to your own, as to both indoor and outdoor noise level, before you place your order. And, if its owner or a salesman is demonstrating it, ask him to start it at its lowest speed—then move up to HIGH. If you follow the reverse sequence you will think the unit is quieter on LOW than it really is. Additionally, turn the thermostat knob so as to start and stop the compressor and check the change in noise level. Room coolers are, of course, subject to Laube's Law of Sound Response: *The hotter the weather the more noise you can stand.*

Here are a few more points I would check on, before signing up for a room cooler:

1. Some apartment houses and office buildings have restrictions against their installation. Check first.

2. If you plan through-wall installation be sure the unit has no side louvers. Or, if it does, that they won't be blocked.

3. If it's going on the sill, can the window still be opened? Washed? (Probably neither.)

4. The cord has a 3-prong plug. Will your receptacle take it?

5. Will your wife understand the controls or are they the super-bewildering kind?

6. Is it going into a space that faces south and is now too warm in cold weather? AHAM standards require the unit not to ice up at or above 70° and 57% rh, outdoors. What happens if you run it on a 50° day? (Page 211).

7. Check the ventilation damper. Will it close tightly enough to prevent cold drafts in winter? The ventilation feature on most window units is now a bad joke. Carrier's first, the 1936 model, did a better ventilating job than any on the market today.

8. Are the air discharge grilles adjustable, so the air won't go where you don't want it to go?

9. Will condensate dripping from the outdoor end of the unit bother you? Most units evaporate condensate over the hot condenser—but not all of it!

10. Is the unit *chassis* construction? If so, the installation need not be disturbed, to pull out the mechanism for inspection or service.

11. Did you check access to the air filter and how you clean it?

12. Does the *5-year warranty* cover just the compressor or the entire hermetically-sealed refrigeration circuit? Does it cover parts, labor, transportation, on-the-job trouble diagnosis?

13. Are you sure you will get a tight, weatherproof installation and at least one year of competent, prompt, maintenance service? (If you buy a unit with a do-it-yourself installation kit, better remove it in the fall and reinstall it in the spring.)

14. How much will it cost? To save money, watch for special prices advertised in January, February or March, or wait until August or September. But why wait? Room coolers are one of the few things that today cost less than they did 20 years ago.

Don't expect to be fully satisfied on all these points. But, at least, give it a try. On a hot day, the ubiquitous window shaker can be a great comfort to you and your family. And your dog, too. At least it was to a pair of pampered canines in Washington. They lived in a room of the palatial home of the owner of the fabulous Hope Diamond—Evelyn Walsh McLean. Their's was a console model and it caused them trouble.

Every morning Mrs. McLean found a little puddle at one corner of the air-conditioner. She cursed and beat the dogs as only a miner's daughter could. She'd teach them not to mistake the conditioner for a fire plug! Then, one day, a serviceman showed up. "What do you want?" she asked, "the unit works okay." "Maybe so," he said, "but the factory made a batch that weren't properly sealed. I've come to fix yours, so condensate won't leak on the floor."

Chapter 10

The Place Called Home

When selecting a new home, the last thing a woman looks at and the first thing she later complains about is the comfort system. Yet it is much more important to her and her family than other features to which she may give far greater attention. Perhaps there is a good reason for this. Comfort, but more particularly the lack thereof, is obvious to her. The mechanical components and functioning of the comfort system—except for adjusting the thermostat, which she is apt to do improperly—are not.

There was a time, in the early days of forced warm air heating systems, when the salesman said: "Cool air from the basement will cool your house in summer." We don't hear that anymore. Why not? Let's look at a few figures: Assume the basement has a floor area of 1250 square feet with an 8 ft. ceiling. Thus, it holds 10,000 cubic feet of air. If it is at 60°, it weighs 764 pounds. How much cooling does it do when it is pumped into the living quarters and warmed to 75°? A total of 2860 Btu's. That is equivalent to a 10,000 Btuh room cooler operating for exactly 17 minutes.

So let us discuss comfort in the home, with emphasis on

summer comfort, and proven ways of providing it. These remarks will apply primarily to the single family house. To a large extent, however, they are applicable also to certain multiple dwellings: those with an individual comfort system for each family. These include most double, row and town houses and many garden apartments. High-rise apartment houses and other multiroom buildings are discussed in Chapter 11.

REQUIREMENTS VARY WIDELY

Taken individually, comfort systems for the home are relatively simple. Taken as a group, they are surprisingly heterogeneous. There are three major reasons for this:

1. Climate—Throughout our country requirements vary widely. This is so, even if we exclude Alaska and Hawaii. Within the borders of 48 states we find deserts and jungles, the arctic and the tropics. We design heating plants for outdoor *design temperatures* of 45° in Key West, for 35° below zero in Williston, North Dakota.

For summer cooling, the range is less. Yet in just one state, California, cooling plants are designed for an average maximum summer temperature of 82° in San Francisco, 110° in El Centro. Assuming 75° indoors, we design for a *temperature difference* of only 7° in one place, 35° or five times as much in another.

2. Economic Aspects—Most of us want, from what we buy, a fair balance between its cost and its utility—the benefits we get out of it. The utility of heating and cooling systems varies widely throughout our country. Take for example, San Francisco. The *amount* of heat needed there is less than half that in St. Paul. But in San Francisco it is needed for more hours per year than in any other major American city—over 2000 hours versus 1900 in St. Paul. What about Miami? There it is needed for only 120 hours during an average year.

A cooling system may be expected to operate an average of only 33 minutes per day, on a year-round basis, if it is in Boston. In Miami the comparable figure is 4.5 hours per day. Thus, the fixed charges on cooling equipment in Boston are eight times as great, per hour of operation, as in Miami.

Is there a break-even location? Yes, Oklahoma City comes pretty close to it. There both the heating and the cooling functions operate about 1100 hours every year. In Boston it's 1800 hours for heating, 200 for cooling. As you would expect, the figures reverse for Miami; 120 hours of heating, 1700 hours of cooling.

When it comes to our homes we are not, usually, too concerned about getting our money's worth. But we should realize that both heating and cooling are a luxury in some areas, a necessity in others.

3. Technical Considerations—It is much easier to decide on a year-round comfort system for a new house than for an old one with a heating system already installed. But both raise questions. Here are typical ones:

    a. My house has hot water baseboard heating. How, best, do I cool it?

    b. Knowing that warm air rises and cool air drops, am I making a mistake when I use the same duct system for both heating and cooling?

    c. Can I save money and get good cooling with a switchover system that cools the living room during the day and the bedrooms during the night? (That is an easy one to dispose of right now. Forget it! To cool both your bedroom and

your bed you will want to switch it over around 7:00 PM.
If you do, you will soon be sweating in front of your TV.
Once the ducts are in, a larger cooling unit costs very
little more.)

  d. If I want to heat the entire house but, to save money, cool
     only part of it, how can this best be done?

Before tackling these questions let me repeat a point already
made elsewhere: do not compromise the quality of the heating
system for the sake of the cooling function in those 43 states
in which heating is always required for more hours per year
than is cooling.

Before proceeding further, let us glance for a moment at the
acceptance of cooling, as an adjunct to heating, in the home.
ARI started compiling statistics on Residential Unitary Air Conditioners for cooling only, or for both cooling and heating, in
1953. That year, member companies shipped 42,000. By 1955
this had increased to 111,000. From then on, shipments have,
on average, more than doubled every five years: 217,000 in
1960, 581,000 in 1965, 1,140,000 in 1970. This does *not*
include room air conditioners.

Your present home is surely heated. If it is not already cooled,
you doubtless want it to be. And, if not, you may be sure your
wife wants at home what you enjoy at work or, in any case, you
both enjoy elsewhere. Having decided to do it, what are your
most reasonable alternatives? Let us consider them:

THE EXISTING HOUSE

Your present heating system is the major factor influencing
the selection of your new cooling system. Construction of the
house comes next. Is it a single or two-story or split-level? Does

it have a full basement, partial or none? Is the attic easily accessible, with enough head room for a man to stand? First, let us assume that you now have a hot water, steam or electric heating system that does a good job and is in good condition. Then we will consider a good forced warm air system. In both cases, what are your reasonable alternatives?

1. House with Hot Water, Steam or Electric Heat

   a. *Hydronic Cooling System:* One thing you *cannot* do is to use a water chiller and pump the chilled water through the heating pipes for summer cooling. Why not? Because, first, your pipes will sweat and drip water which you will soon see coming through the plaster. They are insulated? Not with vapor-proof insulation that will prevent this. But assume they are. Pump chilled water through your radiators, convectors or baseboard heaters and 100 highball glasses, on a humid July night, will sweat less.

   If, however, you can properly insulate the pipes and replace your present radiators, or whatever, with fan-coil units (Chapter 11) you can do it. But it will be a major undertaking. The cost will be high.

   b. *Window Units:* You could do worse. But do not put them on the sill. Install them through the wall. Where? Under or alongside a window. Or, for better air distribution, high on the wall. Put them in only those rooms you use most. With doors open, the others will get some benefit. The installation must be weatherproof. The units should rest on resilient, vibration-absorbing material or else compressor noise, that you don't notice at the unit, may travel along the wall and be bothersome elsewhere.

Room air conditioners give you individual control. And any you do not need may be turned off, for minimum operating cost.

c. *Central Cooling Systems:* This, in one form or another, is usually the best bet for residences, since duct runs are short, air velocities low. This reduces *pumping* power. If you have a two-story house consider using two: one in the attic, to serve the upstairs only. Simple ducts should run along the attic floor to ceiling diffusers. A return air connection will be needed. It should probably run from the ceiling of the hall back to the unit. The hot discharge air from the air-cooled condenser should be ducted through a window or ventilating louver and discharged outdoors. The outside air drawn into the attic by the condenser fan will lower the attic temperature, thus reduce the bedroom cooling load.

The second unit should be in the basement or garage. A main duct along the basement ceiling should have branches going to floor or low sidewall outlets, placed to clear the furniture, preferably under the windows. Or a branch duct could come up through a closet to a high sidewall outlet. The basement unit, if air-cooled equipment is used, should discharge the hot condenser air outdoors. If water-cooled, using city water, you might do what I do: the warmed water leaving the condenser is piped to an outdoor hose connection. We probably have the only yard and garden in the Syracuse area that is sprinkled with warm water.

Two units give you zone control. By turning off the one you don't need, operating cost is reduced. And, if one should quit, the other will still give you a refuge from the heat. Then, too, using this method lets you do as much or as little of the house as you want to.

If you use a single unit for a two-story house you have to duct air between floors, which may not be easy. For a single-story house, with suitable basement or attic space, a single unit would normally be used. ARI-certified single packaged air-cooled units are offered by dozens of companies in the range of 24,000 to 120,000 Btuh (two to ten tons). A lesser number make them water-cooled; usually from 36,000 to 120,000 Btuh. A few go down as low as 18,000 Btuh.

Instead of a single package unit, consider what is now most popular: an air-cooled condensing unit installed on a pad in the backyard, with a direct expansion *air-handling unit* in attic or basement. The indoor noise level will be less.

If you have decentralized electric heat, i.e., baseboard, recessed units or ceiling cable, treat cooling the same as you would for hot water heating.

Always remember that summer cooling is not the reverse of winter heating. In winter we like bedrooms cool and bathrooms warm. But in summer, we don't want the reverse—bathrooms cool and bedrooms warm. And no matter what we do, warm air rises, cool air falls. By using separate systems for both heating and cooling, the design of one need not be compromised for the sake of the other. That is why separate systems usually give the best results.

2. House with Forced Air Furnace—This, in existing houses,

is the situation most commonly encountered. And the most popular procedure for adding cooling is to install a *split* system. What is that? It consists of an air-cooled *condensing unit* located outdoors. This comprises the entire cooling mechanism except for the part that actually does the job—the cooling coil, or evaporator.

a. *Split System:* Today, many furnaces are designed to accommodate an evaporator, often called an *A* coil, because of its shape. In that case, it is installed in the bonnet of the furnace. If there is no space in the bonnet, the *A* coil must be located in the single large duct or plenum through which the air from the furnace travels before it reaches the branch ducts.

Your present duct system presumably does a good job of heating. Before you install a split system or any other for cooling, make sure your duct system, as it now exists, will do a reasonably good job of cooling. If not, what changes can be made in it that will improve the cooling without adversely affecting the heat?

CAUTION—For cooling you should circulate 50% more air than needed for heating. How do you do it? Speed up the blower in your furnace by at least 50%—maybe more. Why more? Because the resistance of your new *A* coil will reduce your normal air flow. What happens when you speed up your blower? Two things: First, your present fan motor will be too small. Say it is rated at one-sixth horsepower. Your new motor will have to be at least one-half horsepower—maybe three-fourths. Fan power goes up as the cube of the speed. A 50% speed increase ups the power 3.37 times. If your one-sixth hp motor is now fully loaded, in theory your new one should be rated at 0.167 x 3.37 or 0.56 hp.

The second thing that happens is that the noise level generated by your blower will increase significantly. That is why most suppliers offer two-speed fan motors. One of the largest has standardized on four-speed motors. Then, when the job is being installed, they select the lowest speed for cooling that will deliver the necessary air. If you find the noise excessive during the heating season, change back to low speed. But do *not run on low while cooling*. If you do, the air will get so cold your ducts will sweat and cause damage.

Finally, some people want to use their old heating thermostats when they add cooling. The combination heating-cooling thermostat has an advantage. Usually it has a three-position switch marked HEAT-OFF-COOL. This can and should be arranged automatically to run your blower on low in the HEAT position; on high when you change to COOL.

    b. The Overgrown Window Unit: Several window units may, of course, be used in houses with warm air heat. Here, however, I refer to something quite different. Particularly in the southwest, many smaller houses are cooled by one single, oversized window unit—one with a capacity in the range of 20,000 to 30,000 Btuh or even larger. How? By running the furnace blower continuously. The unit will, most likely, be installed in the living room. The furnace fan takes air from *all* rooms. The cool air returned from the living room mixes with it. Then it is returned to all other rooms, at a lower temperature. Thus, the oversized unit does not over-

cool the living room. But it does provide cooling for
the rest of the house. It is hard to beat the *comfort per
dollar* provided by this arrangement.

THE NEW HOUSE

As already indicated, you have more freedom of action here
than when you add cooling to an existing house that already has
a good heating system. Basically, however, you have only two
alternatives: do you combine your heating and cooling into one
system or do you use separate and independent systems for each
function? The least costly and most popular method is to use
one combination heating-cooling unit connected to a system of
air ducts serving the entire structure. Let us consider this first:

1. Combination Systems—For heating you can employ gas,
oil, electric resistance elements or an air-to-air heat pump (reverse cycle). The entire cooling mechanism may be built into the
single unit. Or the cooling coil in it may be served by a remote
condensing unit or absorption water chillers (page 161).

Normally, the ducts would be made larger than for heating
only. This makes it easier for the blower to push the necessary
air through them. Result: lower blower speed, less power required, quieter operation. Now the air supply registers will be
selected and located with both heating and cooling in mind.
Possibly the conditioned air will be delivered to the rooms
through high sidewall registers or ceiling diffusers. In that case,
the grilles through which room air is returned to the equipment
must be in or near the floor. In cold climates, however, this
is a poor heating arrangement, unless double glass windows are
used.

If the supply registers are at or in the floor, three precautions
must be observed: *a.* Low wall registers should be capable of

delivering the cool air in an upward direction and for good heating should, as already indicated, be under the windows. *b*. They must be located so that sofas and other furniture will not block the up-flowing cool air. For heating this is less critical —it will rise anyway. *c*. Contrary to the best heating practice, the return air grille should be at least seven feet above the floor. Else, in summer, stratification will give you cold feet and a warm head.

Before deciding on this system keep one thing in mind: you will have to rebalance your air distribution—adjust your registers —twice a year. Why? Your southwest bedroom needs less air for heating, more for cooling than does the one on the northeast, if they are both the same size. Then, too, most people like to overheat bathrooms in winter. But they do not want to overcool them in summer. For bathrooms, my choice is individual electric heaters.

2. Separate Systems—These cost more. They are, however, capable of giving better results. And they broaden your choice of both heating and cooling equipment. You can heat with a separate warm air system, circulate hot water, or use baseboard or other forms of decentralized electric heat. You can even use an electric furnace, though you probably wouldn't. And you won't have to compromise with good heating practice nor with good cooling practice either.

As for the cooling, my recommendation would be the same as that given under The Existing House, Item No. 1 (c). If the house is large, I would consider two separate cooling units, even if single-story: one for the living quarters, the other for the bedrooms.

However, if you have a large house and really want to blow yourself, you might consider using a central water chiller serving individual, thermostatically controlled fan-coil units in each room. If you use under-the-window units, you can get them with built-in electric heating elements. At that location, they will do an excellent heating job. But if you select ceiling mounted units, located above furred down closet ceilings, do not use them for heating unless you are in a mild climate or have double-glass windows.

Let me emphasize that, in the foregoing, I have limited myself to what I consider to be the most sensible approaches to year-round comfort in the home. If every combination and permutation of the available equipment were to be considered, a separate book would scarcely be large enough to cover them.

EQUIPMENT

Furnaces, combination year-round air-conditioners, remote condensing units, *A* coils, air handlers served by remote condensing units and other equipment have already been mentioned. Your attention should, however, also be directed to several rather specialized products:

1. Outdoor Cooling Units—Several companies offer a range of complete, factory assembled cooling units that are intended to be installed through the wall of the house. They have air-cooled condensers and are designed to deliver the cool air through ducts. They come in a range of sizes suitable for all but the largest houses. These, however, could be served by two or more units. Some are also offered as reverse cycle heat pumps. Overall dimensions are on the order of 30 to 40″ wide, 20 to 25″ high and 24 to 30″ deep.

Another version is tall and thin, for *plastering* on the outside of the wall. Thickness runs from 13 to 15″, width 25 to 40″

and height over 6 ft. Duct connections to the house are made through the wall.

2. Apartment Units—Several companies manufacture completely self-contained heating-cooling units suited particularly for low cost apartments and small houses. These combine gas heating with a built-in air-cooled refrigeration system. Installed in a utility room, they are designed to fit flush against an outside wall. Condenser air enters and leaves through a wall opening. Ducts deliver the heated or cooled air to the individual rooms.

3. Absorption Systems—Two types that both heat and cool by the use of gas are on the market. Fans and pumps are, of course, driven electrically. One type is completely self-contained. It employs a solution of lithium bromide in water. The water acts as the refrigerant. This means it is at an extremely low pressure—meaning high vacuum, high enough so the water boils at about 45°. The other type is a split-system. The cooling portion is outdoors. It uses ammonia to chill water. This is then circulated through the cooling coil of the combination heating-cooling unit indoors.

4. Air-to-Air Heat Pumps—These heat by pumping Btu's from the cold outdoor air into the house to warm it. Get heat from cold outdoor air? Yes, all the way down to what is called *absolute zero* which is 459° below zero, Fahrenheit. Example: The outside air is 40° F. The heat pump removes heat by cooling it to say, 20° F. This heat is pumped into the house. To it is added the energy needed to operate the compressor, which does the pumping.

Using these temperatures, assume we have a certain three-ton unit that pumps 24,000 Btuh out of the 40° outside air, thus cooling it to 20°. Its compressor motor draws almost 3000 watts. This, multiplied by 3.4 Btu's per watt is equivalent to another 10,000 Btu. The total quantity of heat delivered to the house is then 24,000 plus 10,000 or 34,000 Btu.

In summer, the heat pump functions as an ordinary cooling unit. Room air is cooled by circulating it through the evaporator coil which, in this case, is called the *inside coil.* Outdoor air is circulated through the condenser, now called the *outside coil,* to cool it. For heating in winter, a *changeover valve* reverses the compressor connections. Then, the inside coil becomes the condenser, the outside coil the evaporator.

In the above example, the heat pump cooled outside air from 40° to 20°. This caused humidity to condense on the tubes and fins of the outside coil. At 32° F water freezes. Thus *icing-up* of the outside coil soon stopped the air flow. This caused the heating also to stop. And this illustrates one of the two basic weaknesses of the air-to-air heat pump. Various automatic gadgets are used to melt the ice and restart the heating, until it is again stopped by ice.

Today, because of the icing problem, most air-to-air heat pumps automatically switch off the compressor and use electric resistance elements for heat below 40° F outdoors.

The second weakness of the heat pump is this: the more heat you need the less you get; the less you need, the more you get. Our three-ton model delivers 34,000 Btuh when it is 40° outdoors. Let's say this is exactly what is needed for comfort in the house at that outdoor temperature. Then, at 20° outdoors, the heat needed to provide comfort increases to 44,000 Btuh. But, at this temperature, the air-to-air heat pump will deliver only about 24,000 Btuh. Conversely, at 60° outdoors only 11,300 Btuh would be needed for heating the house, yet the unit would deliver 47,000. In short, one might properly say that the capacity capabilities of the air-to-air heat pump are bass-ackwards.

No single heating device has had the failure record or the service costs or has caused the disappointments or has been such an all-round frustration as the air-to-air heat pump. One branch of our armed services installed 10,000. Annual failure rate, initially, was 22 to 25%. Finally, they got it down to 10%. To correct each failure cost $400 to $500. Since then, several manufacturers have solved at least some of the problems that have beset the air-to-air heat pump during these past twenty years or so. If you live on the Gulf Coast or in Florida, it is something to consider. But, if you plan to buy one, ask for a price that includes a three, four, or five year maintenance contract—parts, labor and freight.

5. *Ducts*—Steel ducts, even if galvanized, eventually rust if they get wet through condensation or malfunctioning of the equipment. Aluminum ducts cannot rust and are quieter. Ducts in attics and other hot spaces must be insulated or you lose heat in winter, cooling in summer.

Insulation on the outside of ducts will not absorb sound made by the equipment. But on the inside of the duct it will. That, therefore, is where it should be put. WARNING: If ducts are made of fiberglass, edges must be sealed so fibers won't loosen and carry into the living space.

Much progress has been made in manufacturing the entire duct out of insulating material—principally fiberglass. Subject to the warning above, they have a lot of advantages and should be seriously considered, particularly for attics. So should the use of the round, flexible ducts made of fabric and insulating material.

6. Registers—Recently I encountered *tip-out* wall registers. These are for installation at or near the floor. During the heating season, they do not project from wall or baseboard. But, for cooling, they may be tipped out. Then they discharge air in an upward direction. If unobstructed, they can possibly do this with enough velocity to carry the cool air stream up to the ceiling. If so, it may be safe to ignore my previous recommendation to the effect that, if registers are near the floor and used for cooling, the return air grille should be near the ceiling.

7. Air-Cooled versus Water-Cooled Condensers—Do not use water-cooled condensers if local ordinances or water costs require you to recool it by circulating it over a cooling tower. Cooling towers are too much of a nuisance to put up with, if you do not have to. However, in some areas, city water is plentiful and cheap. And it may be used for condensing purposes, for equipment of relatively small capacity.

If this is your situation, you should know that water-cooled equipment does have a few advantages over air-cooled: since you have no condenser fan, it is usually quieter; since the water-cooled condenser can be indoors, completely self-contained equipment is more feasible than is true with air-cooled condensers; since you do not have the cost of running a cooling tower, the operating cost of your cooling equipment will be slightly less. Then, too, the water is not contaminated in any way. Thus you can use it for garden sprinkling or for a swimming pool, if you want it heated.

8. To Split or Not to Split—As I have said before, a hermetically sealed refrigeration system that is completely assembled and tested at the factory has a far lower failure rate than do those that are assembled in the field. A split-system must, of necessity, have the cooling coil—evaporator—connected to the outdoor condensing unit, on the job. To be good, the inside of a cooling system must be clean and dry—really. And it must be tight—with a vengeance—else the refrigerant will be lost. Air always contains moisture. To prevent this from getting into a split system during installation, precharged refrigerant tubing, with quick coupling fittings, should be insisted on for connecting the two parts.

Other things being equal, I would prefer the fully factory assembled system to any split system.

WHAT SIZE DO I NEED?

The first rule to remember is this: for cooling, get the smallest system that is large enough. Take advantage of *flywheel effect* (page 18). Set your thermostat at 70 to 72°. Sleep at that temperature. Let the house warm up to 75° or more during the very hottest part of the afternoon, on the hottest day.

Remember, your thermostat starts and stops the compressor, as required, to maintain the temperature for which it is set. But when the compressor is stopped, you get no dehumidifying. Actually, when the compressor stops, the condensate hanging on your *A* coil reevaporates into the air stream and raises indoor humidity. The smallest system that is large enough gives you far better humidity control than one that is oversized.

Most residential cooling systems have a capacity in the range of 36,000 to 60,000 Btuh (three to five tons). A common rule of thumb for residences is to figure 450 square feet of floor area per ton. However, this can be misleading. Two years ago an air-conditioning contractor friend of mine built a new house near ours. He stopped in the other evening. "How large is your new house," I asked. "About 2100 square feet," he said. "Guess

you have a 5-ton machine," says I. "No, it's just half of that—
2½ tons—over 800 square feet per ton."

I questioned its adequacy. He insisted that he and his family
were very well pleased during the past unusually hot summer.
The answer? His two-story house is very well insulated, has
double glass windows. It faces north. Tall maples constantly
shade the other three sides. That was the answer.

So, whoever sizes the cooling system for your house, be sure
he knows his stuff and does it carefully. I recommend the use
of ARI's Residential Air-Conditioning Load Calculating Forms
for this purpose.

SUGGESTIONS AND REMINDERS

Whatever you decide to do about year-round, home-side
comfort, here are a few additional points to keep in mind:

1. Your cooling coil will remove humidity from the air. This
condensate must go somewhere. So remember the drain pipe.
Incidentally, it is distilled water—good for your wife's electric
steam iron or your car's battery.

2. Insist on having the equipment placed for easy accessibility,
for filter changing and servicing. Mark your calendar to clean
or change or at least inspect your filters at least on January 1
and July 1 of each year. For air filter information see Air Cleaning, page 52.

3. Air from your system can not get into a room unless it
can get out again. If the room lacks an air return grille, undercut
the door at least one inch or get a door with a built-in return air
louver.

4. The capacities of the *A* coil, or air handling unit indoors,
must be matched to that of the outdoor condensing unit.

5. Will your neighbors stand still for your outside noise level? Will it satisfy your code—present or potential? (Page 84.)

6. Only you can decide if you want constant or intermittent air circulation. Every time your thermostat calls for heating or cooling your fan can start; then stop when it is satisfied. Or the fan can run continuously. That is the school I belong to, for these reasons:

   a. Noise is less annoying because the noise level doesn't change.

   b. Temperature control is better, since moving air reduces stratification and, depending on its location, may make the thermostat more responsive.

7. Black roofs absorb 50% more sun heat than white ones.

8. Attics should always have louvered vents or exhaust fans.

9. Insulation and double glass cost more but they save heat, save cooling, permit the use of smaller, lower priced equipment and provide more comfort.

10. If wintertime humidity in an electrically heated house gets too high—and sometimes it does—crack a window and let in a little outside air.

11. Use ARI certified equipment.

12. Whatever you do, try to keep your system simple.

If you are building or buying a new house, you or your builder must answer the question: will oil, gas or electricity be used for heat? Local conditions, and the extent to which you use insulation and double glass, will influence your decision. Nationally, the latest figures from the U. S. Dept. of Commerce indicate these trends for new, single-family dwellings: 11% were heated by oil in 1968, 9% in 1969; gas dropped from 65 to 64% during this period; while electricity increased from 22 to 25%.

As already shown by the growth figures, the acceptance of

year-round comfort systems for the home is accelerating rapidly. In an effort to ignite this market, such systems were offered long before the public was ready for them. Their period of evolution has, therefore, been much longer than their period of acceptance. What does this mean to you? If you buy a system from a long experienced, reliable source, able and willing to give prompt service, you are virtually certain to get a good job.

Chapter 11

The Paying Guest, Tenant, Patient or Pupil

Remember, back in Chapter 1, Arthur Godfrey's outburst; Wayne Parrish's predicament? One was referring to his office, the other his hotel room. One space was for daytime use, the other for nighttime use. One for working, the other for resting. Quite dissimilar applications, aren't they? Not really, as we shall see.

Half of today's entire comfort industry is devoted to supplying equipment and systems for one broad class of structure— the *multiroom building*. These include, principally, apartment houses, college dormitories, hospitals, hotels, motels, nursing homes, office buildings and, on the fringe of this group, primary and secondary schools.

Soon after the use of group therapy air-conditioning got well under way, the occupants of apartment houses, hotels and office buildings were demanding the same summer comfort where they lived and worked as they were learning to enjoy at the movies. Thus was born the demand for multiroom air-conditioning. In 1931, central systems were installed in new office buildings in Fresno and San Antonio. Another such system was installed in a large new Washington apartment project in 1939. But it was not until after World War II that multiroom air-conditioning really got going. And as it grew, so did the complaints about manufactured discomfort.

MULTIROOM REQUIREMENTS ARE DIFFERENT

As we have seen, group therapy air-conditioning usually serves one densely occupied space. Of necessity, it is limited to maintaining an average comfort condition for all occupants. In contrast, the typical multiroom application consists of a multiplicity of largely independent spaces, sparsely occupied. Each space usually has one principal occupant—the boss in his office, his wife in their apartment. And, as we have already seen, average comfort conditions are not for the boss. He knows what he wants. And, since he or his company is paying for the space he occupies, he is determined to get it. But all too often he has to settle for less—much less.

Yet, this is the one area of the comfort business where it is possible to satisfy the individual occupant. This is where the comfort system could cater to and satisfy the too-hots, the too-colds and those in between. Why isn't it done? Because, when the multiroom market was first recognized, the air-conditioning establishment started at the wrong end. It took the available equipment and, initially, available systems and tried to adapt them to satisfy multiroom requirements. They are still trying to cajole the basic central system into doing what, inherently, it is not suited for. In contrast, the developers of the newer systems reversed this procedure. They started with the occupant and worked their way back to systems and equipment which would do easily and naturally what traditional comfort systems do poorly or, if they do it well, at excessive cost. In short, they developed economical systems that do satisfy the occupant. What, then, does it take to satisfy the principal occupant—to assure his full time comfort? Three things:

1. The system must operate during all hours of occupancy.
2. In almost all parts of our country, the system must offer a choice of heating or cooling during all hours of operation.
3. The principal occupant must have complete personal control of his environment whenever he is in the conditioned space.

There is a fourth requirement. It applies particularly to office buildings and schools: If the system is turned off at quitting time—and it usually is—it must be turned on again before the normal starting time. This must be done early enough in the morning to assure comfortable conditions for the occupant when he arrives.

There are those who question the need for having heating

CITY	JAN.-FEB. MAX.	JAN.-FEB. MIN.	JULY-AUG. MAX.	JULY-AUG. MIN.
Atlanta	79	5	103	59
Baltimore	76	−4	102	48
Birmingham	81	3	106	51
Charleston, W. V.	79	−5	102	46
Cleveland	73	−9	103	44
Denver	76	−30	104	41
Indianapolis	72	−19	104	42
Kansas City	76	−13	113	49
Los Angeles	90	28	103	53
Nashville	78	−15	107	47
New York	73	−14	102	51
Oklahoma City	81	−4	107	51

Table 11-1—Recorded Temperature Extremes, Degrees F.[13]

available in summer, cooling in winter. This attitude inevitably leads to a compromise with comfort—at the expense of the occupant. A brief study of Table 11-1 shows, for twelve typical areas, why heating may be needed in summer; cooling in winter.

This is not to say that cooling is *generally* needed when the outdoor temperature during the winter thaw rises to 72 or 73°. But anyone with an office along the south face of a modern glass-walled building is going to need cooling on a sunny day, even at outside temperatures below 50°. I recall the chief engineer of Rockefeller Center telling me that the south facade of the Time-Life Building requires more cooling in late autumn than in July.

THE THREE APPROACHES

Comfort systems may, for convenience, be classified under three broad types, thus:

1. Central Systems—These are systems that convert fuel or electrical energy into heating and cooling effect at a central point.Where combustible fuels are used for heating, it can and is done best in this way. This also applies to cooling done with combustible fuels. Examples: Gas or oil engine-driven compressors; steam turbine-driven compressors; absorption water chillers that use steam in summer from the same boiler that provides heat in winter.

What is known as a *district generating station* is a central plant. However, instead of being located within the building being served, it is located in a separate building and serves not one but several other buildings. To do this, it pumps chilled water and hot water (or steam) from the generating station to all buildings being served. Some of these may be miles away.

2. Semi-central Systems—If a steam-heated building has window units throughout, for cooling, it has, in effect, combined central heating with decentralized cooling for year-round comfort. This makes one version of a semi-central system, though not a good one. But this does make an excellent system if, instead of window units, *packaged terminal air-conditioners* (PTAC's) are used instead. These replace steam or hot water radiators. Pipes from the central heating plant supply warmth. Coolth is supplied by individual built-in cooling mechanisms, air-cooled.

Another form of semi-central system is the reverse of this. Cooling of each space is done by means of chilled water, from a central source. The fan-coil units or other equipment it supplies, are fitted with electric resistance elements. Thus, central cooling is combined with decentralized heating.

3. Decentralized Systems—These are feasible only where electricity is used both for heating and cooling. In simplest form, they consist of a multiplicity of PTAC's. Instead of fin-coils for steam or hot water, they have electric resistance heating elements. Thus, each one is a complete, year-round comfort system, under local control. Each may operate independently from all the others in the building. If desired, they can all be operated as a group from a central point. In either case, they need not operate in the same mode. Some may be supplying coolth while others are supplying warmth, depending on the wishes of the individual occupants.

Having defined the need and explained the three approaches, why is it so many installations fail to satisfy that need? For two reasons: The way they are designed and the way they are operated (page 101). If you know or can find out what kind of a system you have, you can quickly decide what, if anything, can be done to improve your comfort conditions if they are not now satisfactory. The brief descriptions that follow will help you identify your system. It will fit into one of two groups:

TWO-SEASON SYSTEMS

The two most prevalent multiroom systems were never sold as two-season systems. But that is really what they are. They do a fine cooling job in July, and a fine heating job in January. But, if they provide comfort during the changeable weather of spring and fall, it is more a matter of luck than of design. Let us consider them under their proper generic names:

1. Two-Pipe Fan-Coil System—It has the lowest first cost, lowest operating cost, poorest comfort capability and is the simplest of all central systems. Unbelievably, it is still being installed in new buildings today. From the standpoint of the occupant, it has not been improved since introduced, 40 years ago.

In office buildings, fan-coil units are usually seen as cabinets, under the windows. In guest rooms, they may be at the same location or above a dropped or furred-down bathroom ceiling, which creates space between the ceiling and floor above. Chilled water is pumped through a fin-coil in the unit to proved cooling when it is hot; hot water *is pumped through the same pipes* to provide warmth when it is cold. Most fan-coil units have two or three-speed fan motors and are quiet on LOW. They provide all functions of air-conditioning except wintertime humidification and ventilation.

Sometimes you will observe a tiny grille replacing one brick under the windows of an apartment house. This is an outside air connection for such a unit. It is for ventilation. Does it work?

It all depends on the vagaries of the wind and the building *stack-effect*. *Blow-through* of cold outside air in winter may cause cold drafts or a freeze-up, if water circulation stops. Today, the usual practice is to supply a separate ventilating system for fan-coil installations or none at all, except for the bathroom exhaust. Control is manual, by means of a fan switch only or automatic, by use of a thermostat that starts and stops the fan or the flow of water.

2. *Two-Pipe Induction System*—That's the one in the Time-Life Building mentioned on page 2. First cost, operating cost and comfort capabilities of this central system are all higher than for the previous system. As before, the unit is usually under the window but it may be above a furred-down ceiling. Instead of a motor-driven fan, room air is circulated by a multitude of tiny *induction nozzles*. To do this, a fixed quantity of preconditioned ventilating air is delivered at high velocity (up to 45 miles per hour) through circular *conduits* from a central source to the *nozzle plate* in each unit. Escaping through the nozzles, it induces room air through a fin-coil similar to the one in a fan-coil unit. Typically, three cubic feet of recirculated room air join each cubic foot of ventilating or *primary* air. This mixture is then discharged into the room. The primary air is dehumidified in summer, may be, but is often not, humidified in winter. Thus, this system provides all functions of air-conditioning except filtering of the recirculated air. Lint screens are, however, often used.

As contemplated by Dr. Carrier, who invented it thirty years ago, primary air would be warm in summer, cold in winter—directly opposite the temperature of the water being circulated through the coil. A person wanting warmth during the cooling season would turn off the chilled water and be warmed by the ventilating air. Contrariwise, the person wanting coolth during the heating season would turn off the hot water. In actual practice, however, the amount of primary air circulated is so little that only token heating or cooling, or neither, may be available from it. Operating cost is less if the primary air is cool in summer, warm in winter, the same as the water. Then it helps, not hinders, what the chilled water is doing. Thus, it permits the use of smaller, less costly equipment.

Occupant control may be by means of a hand valve or a thermostatically operated valve or damper that bypasses the room air around the water coil. Tenants and hotel guests sometimes complain because they can not locally turn off this system, when not wanted. The primary air, often with a hissing sound, keeps coming.

Both these systems perplex and frustrate the conscientious operator. In Denver, on a cold day in May, one told me—when I asked for some warmth—"I've switched this system from heating to cooling and back again five times this week. And still I'm wrong." One of his problems was that some rooms faced north and were too cool, some faced west and were overheated by the afternoon sun. Another was the day-to-day unpredictability of the weather in spring and fall. Then, too, in summer Denver has an average daily range, as between day and night, of 28°. Often you need heat in the morning, cooling in the afternoon.

Then there is that office building in Chicago. Their telephone operator keeps the complaint score during the first hour every morning. Depending on who's ahead, the too-hots or the too-colds, she tells the operating technician whether to circulate chilled or hot water for the rest of the day.

In short, any system that heats with hot water and cools by circulating chilled water through the same pipes gives all occupants a choice of heating or nothing; cooling or nothing. No such system can possibly provide comfort for all occupants, particularly during the changeable weather of spring and fall. This is true, no matter how carefully it is operated. This is still true despite heavily promoted improvements in air flow that have recently been made. If you are so unfortunate as to depend on such a system for your own comfort, about all you can do, when it doesn't give you what you want, is to complain. It may hasten the day when they modernize the system. And it may make you feel better.

Some years ago, tenants of a New York City apartment house at 200 East End Avenue claimed that the comfort system did not deliver what was promised in the GE folders they had received at the time they were signing up. Tenants on the north side wanted heating; those on the south, cooling, at the same time. The owner sued GE and won a verdict of $129,433. The appeals court reversed the verdict on the grounds that Sol G. Atlas, president of the Corporation, was familiar with the system's limitations before it was installed. It was a two-pipe fan-coil system. No concept has done as much to discredit the air-conditioning industry.

FOUR-SEASON SYSTEMS

While no two-pipe system has the design capability of providing full time comfort, other central systems do. The choice of these is nothing short of bewildering. Those described here are the most prevalent:

1. Three and Four-Pipe Systems—These are hybrid versions of the two-pipe systems just described. The four-pipe version has two independent piping systems, each with supply and return. One circulates hot water, the other chilled water.

In the three-pipe version, hot and cold pipes are also brought to the individual units. To save money, a common pipe is used to return the water from the units to the central system. Owners report that operating costs are usually excessive. Chilled water, instead of having to be cooled from 55 to 45°, may have to be cooled from 75 to 45°; warm water, instead of being reheated from 110 to, say, 130°, may have to be heated from 75 to 130°.

Local controls are available to the individual occupant. These give him a choice of heating, cooling, or neither at all times that hot and chilled water are *both* being circulated. Otherwise their functions are, respectively, the same as two-pipe fan-coil and induction systems.

2. *The Dual Duct System*—The *dual* or *double duct* system is called an all-air system. However, chilled and hot water are used at a central point to cool and heat it. Hot and cold air are delivered to the occupied spaces through separate ducts. Each separately controlled room or zone is served by an acoustically treated *mixing box* supplied by branches from the main ducts. A mixing box serving one room, and under the control of its principal occupant, is more satisfactory than one serving an entire zone consisting of several rooms. Each mixing box, which may be of the induction type, has adjustable dampers—usually thermostatically controlled. These mix hot and cold air in the proportion required to maintain the desired room temperature. Mixing boxes are located under windows (preferable in cold climates) or above furred-down ceilings. This system has the design capability of satisfying all the functions of complete air-conditioning.

3. *The Multizone System*—The *multizone* system is a close, all-air relative to the dual duct system. Typically, it consists of one or a group of multizone units. These are made in capacities

from ten to fifty tons of cooling, plus heating as required, in rooftop or in-house design. The rooftop version usually has a direct gas-fired air heater and built-in, air-cooled refrigeration system. Hot and chilled water from a central source may, however, be used instead. Heated air is supplied to the *hot deck,* cool air to the *cold deck.* These serve a number of zones—usually five to fifteen. A single duct runs from the unit to each zone. A pair of dampers is located where each duct connects to the unit—one for hot air, the other for cold. Each pair of dampers is connected together. As one opens, the other closes. They are thermostatically controlled. Thus the temperature of the air supplied to the zone—preferably an individual room—is such as to maintain the desired temperature. Wintertime humidification is usually not available.

4. *Terminal Reheat Systems*—Again, these are all-air systems. Above a certain outdoor air temperature, all air is cooled and dehumidified at a central point by means of chilled water. For ventilation, some of it is outside air, the same as with dual duct and multizone systems. A single duct delivers this cooled air to *terminal reheat units*—preferably one in each space served. These, too, are often of the induction type, preferably located under the windows in cold climates. In some systems, the quantity of air to each space is kept constant, once the system has been balanced. In others, the air quantity is automatically reduced as the space gets cooler; increased as it gets warmer.

Usually the terminal reheat unit is equipped with a hot water coil but sometimes with electric resistance heating elements. Since cold air is supplied to all spaces, some spaces will be too cool, even in summer, at least some of the time. Thus heating must be available at all times, to counteract excess cooling.

The semi-central and fully decentralized systems have already been described near the beginning of this Chapter under *The Three Approaches*. They were specifically developed for multiroom buildings. The first of these succeeded in partially breaking the shackles that have chained the multiroom air-conditioning business to the central system concept. The second has gone all the way and completely broken these shackles. A further discussion of the fully decentralized system appears in Chapter 12.

HEAT RECOVERY SYSTEMS

In autumn and winter, when the sun hangs low in the sky, the south side of a building, particularly if it has lots of glass, may need cooling while the north side needs heating. The same relationship may exist in the morning between east and west; in the afternoon between west and east. But this applies only to the *perimeter* of the building. The *core* of a large office building, for example, always needs cooling, never heating. The heat from people, lights and office machines must be removed. To heat a space, we pump Btu's into it; to cool it, we pump them out.

Usual practice is to buy the Btu's we need for heating in the form of coal, gas, oil or electricity. Conversely, the ones we pump out of the building, to cool it, we reject to the outdoors; we waste them. But, as shown, both heating and cooling are required simultaneously much of the time. Why not, then, pump the Btu's we remove from spaces that need cooling to the spaces that need heating, to save fuel? That is exactly what a heat recovery system is designed to do. There are many versions of central systems of this type. Office buildings are their most popular habitat.

The vested interests of the lighting industry have promoted

constantly higher lighting levels. This has pushed lighting loads higher and higher. Thus, more and more heat must be gotten rid of, especially from the core of the building. If pumped to the perimeter, this heat may supply part or even all of what is required in winter. When this is done, the building may require no other source of heat when the lights are on, even in the north. Hence the catchy promotional expression: *heat of light*. As if there ever was a time when light did not produce heat! The sun does it. So does a candle.

Not only are high lighting levels good for the lighting industry, they are also good for the cooling industry. More light, more heat; more heat means not only a longer cooling season but larger cooling equipment. To cut the cost of summer cooling in heavily lighted buildings, some ingenious devices have been born: Fluorescent fixtures get so hot that air and water-cooled versions are now made. The latter may be cooled directly by cooling tower water. Air-cooled venetian blinds have also been developed. These, when exposed to direct sunlight, may also be cooled by water from the cooling tower.Thus, the load on the cooling equipment is reduced.

Another way of conserving energy by redistributing heat is confined to the ventilating air. It may be used with a wide variety of comfort systems—not just the so-called heat recovery versions. In winter, warm, stale air exhausted from the building is used to warm cold, incoming fresh air. Conversely, in summer, cool exhaust air helps to cool hot incoming fresh air.

With this background, let us now consider the two basic types of the same concept: the heat recovery, heat redistribution or energy conservation systems.

1. Central Heat Recovery Systems—Any central system that is designed simultaneously to provide both heating and cooling may be designed for heat recovery. This includes all of the 4-season systems already discussed. The central refrigeration machine is used simultaneously for both heating and cooling. The heat it removes, by means of chilled water, from spaces that must be cooled is pumped into and heats condenser water. This water is then used to serve those spaces that require heating.

For useful heat recovery to take place, two conditions must exist: (1) A surplus of heat must be available in one part of the structure. (2) A heat deficit must exist in another part. The surplus heat is used to make up the deficit in whole or in part. If there is a net surplus of heat, a cooling tower must be used to get rid of it. This always occurs in summer. If there is a net deficit, as usually happens in winter, it must be made up by the use of a fuel-fired boiler or electric heat.

Some unbelievably complex engineering ecstasies have been developed in the form of central heat recovery systems. Their first cost has been outrageous. It takes a Ph.D. from MIT to operate them properly. When their designers refer to *sophistication,* they really mean a proliferation of complication. How are such engineering adventures justified? Presumably, by lower operating costs. Whether this is achieved depends on the application, the system used and how effectively it is operated. Note that the heat recovery feature, as such, neither adds nor detracts from the inherent comfort capabilities of a similar system without heat recovery.

2. The Decentralized Heat Recovery System—Instead of using one or more large central water-cooled refrigeration machines to serve the entire building, this system uses a multiplicity of small, decentralized, water-cooled units, i.e., water-to-air heat pumps. Smaller sizes are designed for under-window and ceiling installation, larger ones are floor mounted.

Cooling capacity usually ranges from 6500 to 60,000 Btuh. They are equipped with the same reversing valves as are air-to-air heat pumps. Each unit can supply either cooling or heating, independently of all others and under local control. All units are connected to one piping system through which water is constantly circulated. The temperature of this water ranges from 65° to 95°. Thus the pipes are not insulated. Machines operating in the cooling mode pump heat into the water. Those supplying heat pump it out of the same water. A cooling tower rejects surplus heat in summer to prevent the circulating water exceeding 95°. To prevent the water from dropping below 65°, a fuel-fired boiler may make up any shortage of heat in winter. The trend, however, is to equip each unit with a built-in electric resistance element to supply the heat whenever the temperature of the circulating water falls below 65°.

Since each unit is under local control, it offers occupants an independent choice of either heating or cooling. Since the water from which heat is pumped during the heating mode doesn't fall below 65°, the headache of the air-to-air heat pump—the freeze-up problem— simply does not exist. Note, however, that the units, as now built, rarely include provision for introducing outside air. Thus, a separate ventilation system is usually needed, as with fan-coil systems.

Where an abundant supply of natural water in the range of 65 to 95° is constantly available, an interesting variation of this system is possible. This variation is popular on both coasts of Florida. There sea or well water is the source from which heat is pumped in winter by the individual units; and the sink into which the excess heat is pumped in summer. Thus, neither a cooling tower nor supplementary source of heat are needed. Caution: Sea water is corrosive, well water often so. Thus a heat interchanger, made of corrosive resistant material, must be interposed between the natural water and the fresh water in the loop being circulated through the units.

WHICH SYSTEM TO SELECT?

Are you building a new multiroom structure? Are you remodeling an old one now defaced with window unit hangover? In either case, the success of your project depends on one word, *occupancy*. Other things being equal, this in turn depends on what the occupant gets for what he pays. Your costs influence what you must charge. Your comfort system influences the satisfaction of your occupants. Thus the more comfort you can supply per dollar, the stronger your competitive position and the more successful your project.

If you are not a builder, you are a renter. It may be for one night in a motel room, one year in an apartment, ten years in an office building. Once you accept the rate, your main concern is the satisfaction you'll get from the space—particularly the comfort system. And you couldn't care less about what that system costs the owner. Or could you? You, and others like you, are going to pay for that system and its operating cost. In the field of multiroom buildings, there is no more fallacious cliché than *you get what you pay for*. You always pay for what you get. But all too often, as Messrs. Godfrey and Parrish suggested, you may pay and not get. Don't get what? The temperature you want when you want it.

So, since full-time comfort is the object of the game, let's see how close you come to getting it—full-time—from the various systems. After that let's see what they cost to install and operate. You may be in for a surprise.

1. Occupant Comfort—As already shown, two-pipe systems lack the ability to provide full-time comfort, no matter how well operated. All the other systems possess that ability, they possess it *as designed*. In the case of central systems, however, they rarely deliver it, *as operated*.

Remember the well-known Chicago office building mentioned on page 11? No matter how hot the weather, tenants working nights or on weekends have to do so in discomfort. Remember the two hotels mentioned on page 101? One had its system operating, but offered me the function that I did not want. The other was off and provided nothing. Is this the exception? No, this is the rule. The exception is the building with the four-season central system that is operated so that it constantly delivers the comfort that was designed into it.

Why don't owners of four-season systems operate them as such? Because, typically, when that character in tight pants and twisted underwear, the chief accountant, discovers the cost of running a boiler in the *cooling season* and a refrigerating machine in cold weather, something happens. He rushes to the boss with the bills in his hand. Quickly, orders are issued: "Stop the heating May 15, start it Oct. 1; stop the cooling October 15th, start it May 1". And then, for eleven months a year, he has nothing better than a two-pipe system.

The situation is usually mitigated in the case of applications, such as the cores of large office buildings, that simply must have cooling in winter. This is most economically done by increasing the amount of outside ventilating air, where the system is designed to do this.

Some years ago, two of us came up independently with the same idea. A senior officer of Arthur D. Little Company, the great research firm, proposed something he called a *hedonic index.* I called the same thing the *comfort factor.* We both urged ASHRAE to adopt the idea. Each different system was to be given its particular comfort factor. This took into consideration the capability of the system, *as designed,* and the degree of comfort it actually provides, *as normally operated.* The first factor was to be determined by analysis, the second by a field survey. Top officials of ASHRAE have now recognized the need for something of this sort.

My comfort factor is defined as the total annual hours of occupancy divided into the number of hours during which the occupant has the temperature he wants, expressed as a percentage. Such comfort factors, for popular multiroom systems, are illustrated in Figure 11-1. The numerical values reflect the way the various systems are actually operated, based on a limited field survey. They are average values and do not necessarily fit any one specific installation.

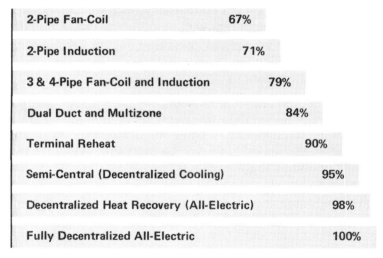

2-Pipe Fan-Coil	67%
2-Pipe Induction	71%
3 & 4-Pipe Fan-Coil and Induction	79%
Dual Duct and Multizone	84%
Terminal Reheat	90%
Semi-Central (Decentralized Cooling)	95%
Decentralized Heat Recovery (All-Electric)	98%
Fully Decentralized All-Electric	100%

Figure 11-1—Comparative Comfort Factors

If you work or live in a building that has a system designed for full-time comfort but don't get it, check your lease. If it limits the hours during which you are entitled to cooling and heating, you are stuck. If not, you have the basis for a valid complaint. And if you are thinking of moving into such a building, read the fine print first. If it is a motel, check the room before you sign the register. They are accustomed to this. And if you need cooling—or heating—and don't get it, refuse to pay the bill. You may get away with it. If not, you are entitled to a discount. As for office buildings, some of them charge as much as $100 per hour to operate their systems at hours other than those shown in the lease. Others simply refuse to do it at all.

2. *First Costs*—To keep it simple, let's consider office buildings only. Today, their comfort systems cost from $3 to over $10 per square foot. Some systems increase the cost of the building and reduce the rentable floor space. Others reduce the building cost and require no rentable floor space. Let's take 8% for mortgage money, 2% for taxes and 1% for insurance. We will amortize the comfort system over 20 years. In that case, a saving in first cost of $3.00 per square foot cuts fixed charges by 31.5¢ per square foot every year. The cost groupings that follow cover the systems discussed.

Low—

　　Two-Pipe Fan-Coil (No separate ventilating system)
　　Fully Decentralized

Medium—

　　Two-Pipe Induction
　　Decentralized Heat Recovery
　　Terminal Reheat
　　Three-Pipe Fan-Coil
　　Multizone (Direct fired—direct expansion)

High—
   Four-Pipe Fan-Coil
   Three and Four-Pipe Induction
   Dual Duct
   Central Heat Recovery (sophisticated!)

The quality of a system has a significant influence on its first cost. A quiet system always costs more than a noisy one of the same type. Whatever central system is selected, its first cost, to the individual building owner, is reduced if he uses hot and chilled water from an independently owned district generating station.

All-air central systems require more floor space or greater floor-to-floor distances than water circulating systems. Semi-central systems take less floor space than either of these. The fully decentralized system takes the least. Systems with cooling towers require roof reinforcing. Fuel-fired systems require chimneys. All-electric systems do not. These are factors that, in varying degrees, influence the cost of the building itself.

   3. Operating Costs—Every July, the annual Office Building Experience Exchange Report is published by Building Owners and Managers International, Chicago. Among other things, this shows, for leading cities, a breakdown of average costs for many items, including heating and cooling. The latest edition, Pages 56 and 57, shows a heating cost in Wilmington, Delaware, of 21.6¢ per square foot per year; in St. Louis of 10.6¢ per square foot per year, even though Wilmington has a slightly milder climate. It shows Cincinnati's cooling costs as 15.3¢ per square foot per year and Philadelphia's, where less cooling is required, as 27¢. Cooling costs as high as 75¢ per square foot per year have been reported elsewhere for systems served by so-called district generating stations.

Obviously, fuel costs and electric rates are not the same in these cities. But they are not as different as these comparative costs indicate. Choice of the comfort system, and how it is operated, has a profound influence on operating costs, hence on rental rates and/or profits.

Operating costs of the various multiroom systems, maintenance expense included, may be grouped in about this order:

Low—

　　Fully Decentralized
　　Semi-central (decentralized cooling)
　　Decentralized Heat Recovery
　　Two-Pipe Fan-Coil

Medium—

　　Semi-central (central cooling)
　　Two-Pipe Induction
　　Four-Pipe Fan-Coil
　　All-Air Multizone
　　Special Heat Recovery Three-Pipe Fan-Coil

High—

　　Four-Pipe Induction
　　Usual Three-Pipe Fan-Coil
　　Usual Three-Pipe Induction
　　Terminal Reheat
　　Dual Duct
　　All systems served by district generating stations

Individual system design, relative energy costs, operating practices and the size of the building all influence operating costs. The foregoing sequence contemplates buildings of five stories or more with total floor areas upwards of 50,000 square feet.

What a surprise we get, what a paradox we observe, when we compare comfort factors, first costs and operating costs. We see that the two systems that provide the least comfort and the most comfort have two things in common: low first cost and low operating cost. Those in between all cost more.

REQUIREMENTS VARY

Basically, any specific comfort system for one application is the same as for another—the same for an apartment house as for an office building. In detail, however, requirements do vary, depending on the application. This does influence the selection of the system for a specific job and details of its design. Briefly the differences are as follows:

1. Apartment Houses—The system selected should not mix return air from several apartments and then return it to the same or others. Some owners include their cost of electricity, gas and water in the rent. Others—especially co-ops and condominiums —want each tenant metered separately. Separate metering of the cost of operating a central comfort system is not feasible. The fully decentralized system and the decentralized heat recovery system are easily metered separately. Then, each tenant pays only for what he uses. It has been observed that the amount of energy used for operating systems that are individually metered runs 25 to 50% less than where a common meter is used and the owner pays the bill.

Rule of Thumb: About 450 square feet per ton of refrigeration.

2. College Dormitories—These are considered to be the same as apartment houses. Special precautions against pilferage and vandalism may, however, be desirable.

3. Hospitals—A doctor with a heart patient may request a room temperature of 65°. One with a patient in diabetic shock may ask for 80° or even higher. The system should be able to satisfy both requirements at all times—one room on cooling, the other on heating. There should be no interchange of air between patient rooms. Circulation between corridors and patient rooms should be minimal. With doors open, the best that can be done is to exhaust from the room or, preferably, adjoining bathroom, as much air as is supplied. Great care must be exercised to design a system that prevents rather than aids the spread of *staph* and other airborne infections. The requirements of offices, laboratories, assembly halls, nurseries, recovery rooms and operating rooms are all different.

Operating rooms should always be on a separate system, with an alternate source of emergency power such as an engine-driven generator, to serve the heavy lighting and cooling loads. Usual practice is to specify a costly 100% outside air system for operating room, to diminish the explosion hazard. Research conducted in Barnes Hospital, St. Louis, indicates this is not necessary. In any case, air-conditioning for hospitals is a highly specialized and controversial subject, one that cannot receive more than cursory attention here. A word of caution is, however, in order.

If Federal money is involved, HEW's Division of Hospital and Medical Facilities must approve the comfort system. But that does *not* mean that their original recommendations must be explicitly followed. Their representatives are sometimes less sensitive to the spending of money for doubtful benefits and complex systems than the taxpayer who, ultimately, must foot the bill.

4. Hotels/Motels—Always use separate systems for guest

rooms and public rooms. All must be served at certain hours, only the guest rooms at others. For guest rooms, the trend is rapidly in the direction of decentralized all-electric equipment. The use of the decentralized heat recovery system is also increasing. But this is more for comfort than for the heat recovery feature. For operating economy, many motels use central control panels, located at the front desk, for either of these systems.

Recently, I checked into such a one. Their control panel had a toggle switch for each room. After I registered, the clerk, in one operation, pulled my key off of its toggle switch and snapped it down. When I checked out, she snapped the switch up and hung my key back on it. What could be simpler or more economical? But, in cold climates, units controlled in this way require an automatic *set-back*. This prevents the room from dropping below its fixed setting—typically 55°—when the controls are in the OFF position.

Where an independent system is used for the guest rooms, indoor package units, rooftop equipment or even a central system is used for lobbies, restaurants, convention halls and other public spaces.

Rule of Thumb: 10,000 to 12,000 Btuh cooling capacity for guest rooms facing west and corner rooms; 9000 for the others. This is where equipment is turned on when the guest arrives. If the equipment runs constantly, reduce this by one-third.

5. Nursing and Retirement Homes—For nursing homes, the requirements are similar to but less exacting than for hospitals. For retirement homes, they resemble those of small apartments and motels. Remember that controls must be as simple as possible. And elderly people prefer higher temperatures than young ones and are more sensitive to drafts.

6. Office Buildings—The typical new office building requires two systems for its rentable office space: one for the perimeter, the other for the core. Sometimes the core system also serves the ground floor and basement. Older office buildings being modernized have been successfully conditioned with perimeter systems only, where no occupied space is over 30 to 35 feet from an outside wall. This recognizes that the core need not be as cool as the perimeter in summer, for three reasons: the occupants are more apt to be women; occupants arriving from off the hot street usually wait in the core before being ushered into a private office on the perimeter; direct or radiant heat from the sun does not reach the core.

The perimeter system must be able to respond quickly to widely varying outdoor conditions: temperature, wind velocity and constantly moving sunshine, clouds and shadows. The core system, except for the top floor and, if served, the ground floor, needs to provide cooling only. There, only one load normally changes: the temperature of the outdoor ventilating air.

Comfort is always a compromise where the ladies, bless them, are concerned. The thermostat in offices occupied by several women should be locked. The key should be in possession of the office manager or some other man in a senior position. With the tact and wisdom of Solomon, he may be able to keep the peace. The manager of a beautiful new building in Boston gave the key to the senior one of three women in a single office. Before the situation calmed down, he supplied a desk fan for No. 2 and a small electric bathroom heater for No. 3.

The rental agent has a competitive advantage if the system selected is able to provide economical comfort for occupants who must work after normal hours or on weekends.

Rule of Thumb: 200 to 350 square feet per ton, depending on location, light intensity and amount of glass in outside walls.

7. Schools—Taxpayers will not go for air-conditioning. But they do favor an environment for learning. Hence the trend is to provide today's primary and secondary schools with environmental control systems—this being a more acceptable expression. Overheating is the problem. Take a standard 30-seat classroom of 750 to 900 square feet, facing south, at 40° latitude (Philadelphia, Indianapolis, Kansas City). Assume 100 square feet of window area, draped. With 30 youngsters squirming in their seats, the teacher trying to control them and the lights on, heating would not be required on a sunny, December day—except for morning warm-up—until the outside temperature dropped below 20°. Where the system is designed for it, outside air can be used for cooling when its temperature is below 60°. Above this, mechanical cooling is needed. The automatic controls for accomplishing this are costly and complex and may entail sizeable service expense. Lack of cooling in warm weather makes pupils drowsy and inattentive, at the expense of take-home learning.

The equipment should be as vandal-proof as reasonably possible. Room thermostats should be under lock and key. Teacher wants the key. If she has it, the room is too warm to suit the kids. That is why many school boards give the key to the custodian.

Five to seven cfm of ventilating air per pupil is a good number for schools. If your applicable code calls for more, try to get a variance. It will save the taxpayers money.

School designers are getting away from the standard 30-pupil (40 or more in parochial schools) classrooms. They are going for wide open spaces with movable sight and sound barriers instead. This makes a modular comfort system—one that individually serves every 300 to 450 square feet—highly desirable.

As already mentioned, when it comes to multiroom buildings, schools are on the border line. Cafeterias, auditoriums and the like are, obviously, densely occupied group therapy applications. But, for that matter, so are the classrooms. For these, the rule of thumb is 3 to 4.5 tons of cooling capacity per standard classroom with outside exposure and windows. For interior rooms it is less.

TOTAL ENERGY

Did you ever hear of He Li Po? No, he is not a Chinese poet of the Ming Dynasty. The expression stands for Heat, Light and Power. Years ago, a company promoted the idea of a He Li Po plant in every house. Each plant consisted of an engine—gas, gasoline or diesel—an electric generator, a heat recovery system and controls. The worthy object was economy, through more efficient use of fuel. Driving an engine, the energy in the fuel breaks down about like this: one-third useful work, one-third jacket water, one-third loss through exhaust gas and radiation.

The engine, connected to the generator, ran continuously. This provided electricity for lights and other household uses. The heat in the jacket water and exhaust would heat the house, supply the domestic hot water and, in a heat-actuated refrigeration machine, could be used for summer air-conditioning.

For years He Li Po seemed dead. Then it resurfaced in giant form—not as He Li Po, but as *total energy*. Over the last fifteen years, several hundred installations have been made. They use gas or diesel engines or gas turbines in the range of 100 to 1000 or more horsepower. Where? Mostly in establishments of companies that have an axe to grind: gas and oil companies and manufacturers of the equipment. And there are indications that some people in Washington are thinking of them for large housing projects, perhaps to confirm Government's knack for finding the most costly solution to any problem.

Total energy jobs are expensive. This can be a bonanza for consulting engineers, equipment suppliers and contractors. Quite a number of those that were bought at arms-length have been replaced. With what? Conventional systems using electric power supplied by the local utility. Why? Operating costs higher than estimated, breakdowns, inability to hire personnel qualified properly to operate such complex systems.

In those manufacturing plants and processes in which all the waste heat can be used at all times, they have shown economies. For comfort systems, their actual economics are highly doubtful. So much for total energy. Now a word about controls.

CONTROLS

User controls almost always include a thermostat. It is *not* a throttle! The farther you push your right foot down in a car, the faster it goes. But a thermostat will not get you faster heating or cooling no matter how far you push or turn it. So listen carefully, as you move it slowly in the direction you want to go. When you hear a slight click or something start operating, you stop! Otherwise, chances are, your room will overheat or overcool. In short, don't over-control. And if the thermostat is in a public space, be sure it's under lock and key.

Some room-side equipment is available with a choice of three-position or two-position controls. The first offers a choice of HEAT-OFF-COOL; the second, simply ON-OFF. Chances are the latter is of the *automatic changeover* type. You set it at the temperature you want. If room temperature drops, heating will start; if it rises, cooling will be turned on.

The automatic changeover control is ideal for a private office.

There, it is normally operated by the same individual. He learns not to over-control. But it has been found to be undesirable for hotel/motel use. It is so simple it confuses the transient occupant. He is better off with HEAT-OFF-COOL. It forces him to decide what he wants, HEAT or COOL. This seems, once he has done it, to give him a certain psychological satisfaction—a sense of achievement.

AUTOMATION

There is no such thing as complete central system automation. As used here, the term refers to centralized observation and operation. A man at a control console observes and controls the operation of all major components of the system.

Complete automation is, in contrast, available for decentralized all-electric comfort systems. It is already in extensive use in office buildings and schools. No operating technician whatever is required. Yet, it economically assures the full-time comfort of occupants who put in overtime.

Typically, in an office building, all equipment automatically starts early enough in the morning to assure comfortable conditions at starting time. Each space—assuming automatic changeover controls are used—then reverts to the temperature for which its thermostat has previously been set.

At quitting time all equipment stops. This starting and stopping normally occurs on Mondays through Fridays only. It is all done by a central black box which sends impulses over the power wires to the individual conditioners. The important point is that the power at each conditioner stays on, even after it is stopped.

Thus anyone working after hours or on weekends simply taps a reset button on his equipment to restart it. After quitting time, the central control sends additional impulses at preset times, such as at 9:00 PM and Midnight. These automatically stop any equipment that has been restarted—but leave the power on for later restarts, if desired.

GETTING WHAT YOU WANT

The entire multiroom building industry has been so thoroughly indoctrinated with the central comfort system concept that superior innovations have attracted about as much ridicule as respect. Since nothing is harder than upsetting long established habits, how then does the owner of a new multiroom building get the system that contributes most to the bottom line of his profit and loss statement because it offers more comfort at lower cost than can competing buildings?

The motel chains dictate what they want, based on experience. Admittedly, however, some have gone too cheap. Some office building developers tell the architect and consulting engineer what they want, based on what they have learned from experience and by investigating the newest developments. One developer I know of does not permit his architect to do the mechanical engineering or engage the consulting engineer. He picks his own consulting engineer and both he and the architect report directly to the developer. To top it off, he then hires an outside air-conditioning consultant to be a member of his design team. This is the man who can ask the engineer those questions which the owner himself would ask, if he knew enough about air-conditioning to know they existed.

As is so often the case, the important thing is to investigate before you invest, whether you are building a building, leasing an office or renting an apartment.

Chapter 12

The Virtues of Decentralization

Despite old habits, there is today a rapid shift toward increasingly decentralized comfort systems for multiroom structures, schools included. The fully decentralized system uses, of necessity, electricity for both heating and cooling. The decentralized heat recovery system increasingly uses electricity for both functions. Accordingly, we should ask: what influence has the choice of energy on the occupant's comfort?

Anything in the comfort field that can be done with gas or oil can also be done with electricity. But the converse of this is not true. Electricity can do what the other energy sources *cannot* do. Being our most highly refined source of energy, it offers a broader choice of alternatives. This is what accounts for the accelerating swing in the direction of *all-electric* or *total-electric* comfort systems, particularly for apartment houses, hospitals, hotels/motels, office buildings, schools and other multiroom structures. And most of these systems are decentralized.

For the moment, therefore, try to supress your conviction that you cannot afford to use electricity for heat—"It costs too much!" Pretend, if you can, that you are located in Tennessee, where extremely low-priced power is supplied by the TVA, at

the expense of taxpayers elsewhere. Then consider these virtues
of decentralization.

BENEFITS TO THE OCCUPANT

While these advantages are of direct benefit to the occupant,
indirectly they also benefit the operator and owner through increased occupant satisfaction:
1. It assures the full-time comfort of the principal occupant.
   a. He is in full control of his environment, since the choice
      of either heating or cooling is immediately and constant-
      ly available at full capacity and is independent of ad-
      joining spaces.
   b. Systems for office buildings and schools may be hand
      operated or easily and inexpensively automated, to
      assure comfortable conditions at starting time in the
      morning.
   c. The same controls that do this permit occupants who
      work late or on weekends to do so in an environment
      of their own choosing, without wasteful operation of
      unneeded equipment.
2. In the case of apartment houses (some office buildings, too)
the equipment may be individually metered so that the occupant
pays only for what he uses.

BENEFITS TO THE OPERATOR

Be he the manager of an office building, the administrator of
a hospital, the custodian of a school or whatever, the operator
of an air-conditioning system would like to avoid complaints.
These are customary with central systems but are minimal for
the decentralized type, since occupants may always have the

conditions they want. But the operator also enjoys these other advantages:

1. Wasteful operation of unneeded equipment in unoccupied spaces can be easily avoided.

2. Since the system is installed in modules, it facilitates moving partitions, as is often required in office buildings and increasingly in schools.

3. It may be easily and inexpensively automated, in which case the cost of operating technicians is saved.

4. Some versions use no water, require no messy cooling tower and thus no bothersome water treatment.

5. Where the system is fully decentralized, malfunctioning of one unit affects only its space. It affects others no more than does the failure of one lamp in a modern string of Christmas tree lights. The failure of any one of dozens of different components in a central system will stop it entirely, much as the burning out of one lamp in an old-fashioned string of Christmas tree lights turns off all of them.

ADVANTAGES TO THE OWNER

In addition to the indirect benefits the owner receives from contented occupants and a carefree operator, the owner who selects a fully decentralized comfort system receives these direct benefits:

1. The fully decentralized system may initially be installed for automatic heating, circulation and air filtering only, to reduce initial investment. The most costly component, the cooling increment, may then be provided at the option of the individual occupant, often on a self-liquidating basis or as money becomes available. This has frequently been done in apartment houses, hospitals and schools and sometimes in office buildings.

2. "Philadelphia Plumbers Get $19,250 Annual Pay Pact"— headlines the Washington Evening Star of April 27, 1969. The need for costly field labor to install these systems is minimal. The number of trades involved may be only one-third those used to install a central system. This reduces greatly the nonproductivity caused by haggling over work jurisdiction. Thus, the installed cost of the decentralized system is several dollars a square foot less than for central systems that have the design capability of providing full-time comfort.

3. These systems usually reduce total building cost in three ways: by saving floor space otherwise required for equipment rooms, by often permitting reduced floor-to-floor distances, and by eliminating the need for a chimney.

4. The major portion of the first cost may usually be written off in 10 years vs. 20 years for central systems.

5. Strange as it may seem, the operating cost of the fully decentralized all-electric system is usually less than for a central year-round comfort system. The larger the building, the more apt this is to be the case. Why? Because of the unmatched efficiency with which it utilizes the electric energy required for its operation. This will presently be discussed in some detail.

In short, the decentralized all-electric comfort system does more with less. In this age of impending scarcities, this may be its greatest virtue.

It is at its best as a perimeter system. It is compatible with any type of air or water-cooled core system, whether also decentralized or otherwise. But it cannot be used where party walls adjoin other buildings or where glass is down to the floor.

And, this is where the all-electric decentralized heat recovery system is at its best. It can be used for any part of a building, whether perimeter or core. Generally, however, its first cost and operating cost are somewhat higher than for the fully decentralized all-electric system. Compared with central systems, however, it is highly economical.

WHAT EVERYBODY KNOWS

In the air-conditioning industry there were, until recently, three things that *everybody knows*. They were believed by all, challenged by none:

1. It costs more to operate air-cooled refrigeration equipment than water-cooled.

2. The expense of maintaining a multiplicity of small machines is greater than for maintaining one central system to serve the same job.

3. "You cannot afford to use electricity for heating—it costs too much."

Let us look at each of these:

1. Water-cooled condensing costs less—One rarely hears this claim any more. Experience has proven it wrong. Where air-cooled condensers are feasible, operating cost is less. But where they are not aesthetically or technically feasible—because of the large air quantities required—water-cooled condensers will continue to be used.

2. Maintenance Expense—Where the decentralized equipment is of high quality, long term maintenance contracts are regularly taken by reliable, experienced service companies. Usually, these run from five to fifteen years. They include parts and labor. Present prices run from $10 to $15 per ton per year. Comparable charges for similar contracts on central systems— when available—regularly run from 30 to 100% more. This relationship is effectively disposing of this second item of conventional air-conditioning wisdom.

3. Electricity Costs too Much—This is easy to prove. Where a large building is supplied by a tax-paying, privately owned electric utility, a rate of 1.5¢ per kwh is not unusual. One kilowatt delivers 3412 Btu's of heat. Figure it out and you will see one million Btu's cost $4.40. What does gas cost? A price of 60¢ per million Btu's is quite common. Often it is less, on an interruptible basis. This means you may have to switch to oil or expensive *bottled* gas during the coldest weather. In any case, at the figures used, electricity for heat costs 7.3 times more than for gas. So, forget it!

But wait a minute! I have before me an unsolicited letter from an experienced developer of office buildings. It is about a new 26-story office tower in a large city on the shores of Lake Erie. The building has been operating three years. It has a gross floor space of 450,306 square feet. An automated, decentralized all-electric system is used for the perimeter; electrically-driven water chillers supply the core by means of an all-air distribution system. All heating is done electrically.

Directly across the street is another new office tower, a year or two older. By coincidence it, too, has 450,000 square feet of floor space. It has an induction system, with gas-fired boilers for heat; electrically-driven water chillers for cooling. Its energy cost last year was 26¢ per square foot for electricity. Gas costs (so far unavailable) are estimated to have been 7¢ per square foot, for heating. Total: 33¢ per square foot. What about the all-electric building? Cost of electricity is 27¢ per square foot per year and nothing for gas. In both cases, this covers heating, cooling, lights, elevators and office equipment.

Since all the electricity that enters a building ends up as heat, it is impossible to isolate heating and cooling costs from the total cost. These figures, however, show a saving in energy cost alone of 6¢ per square foot per year or $27,000 annually for a 450,000 square foot building. The perimeter system of the all-electric building shows additional savings: no water or water treatment, no operating personnel, lower maintenance expense. And it takes up no valuable floor space.

Is this example unusual? No. The first large all-electric office building I have figures on is in the Washington area. Gross floor space: 272,000 square feet. Its perimeter system is decentralized all-electric, about 500 tons of cooling. Its 200-ton core system consists of package units on each floor. It shows a saving of 6.5¢ per square foot per year for lighting, heating and cooling, as against the average Washington costs reported by members of BOMA International in their Office Building Experience Exchange Report.

Do these buildings have double glass or more insulation than normal? Not at all. It is their size that makes insulation unnecessary. It is easy to show, mathematically, that, as buildings get larger, the heat generated internally per square foot of floor space increases much faster than the area of the walls through which, in winter, it is lost. Insulation would shorten the heating season but lengthen the cooling season. It would make a large building more comfortable but savings in operating expense would not amortize its extra first cost.

Chain motels were first to recognize the cost savings and the increased guest satisfaction of the decentralized all-electric comfort system. Available information indicates that over 75% of the new ones built since 1969, or now underway, use this system. Where glass is down to the floor, it is not feasible. There, the decentralized heat recovery system is making headway.

ENERGY USAGE

How, then, is it possible to use energy that costs five to ten times more than gas and still have a lower operating cost? These are the four major reasons that account for this startling paradox:

1. *Temperature Lift*—When a refrigerating system raises the temperature of its refrigerant from 45° in its evaporator to 100° in its condenser, we speak of a temperature lift of 100 minus 45 or 55°. These figures happen to be the average values for a package terminal air conditioner operating in Chicago during the month of July. Concurrently, the refrigerant in the evaporator of a central water chiller is at 30° in order to cool the water to 45°. If its condensing temperature also averages 100° (chances are it will be higher), then its temperature lift becomes 100 minus 30 or 70°. Since, in this comparison, 70° is 21% more than 55°, the central system takes 21% more energy than the decentralized system, for cooling.

2. *Traveling Distance*—The farther fans and pumps push air and water, the more energy it takes. In a typical 20-story office building with a central system, each Btu pushed in for heat in winter and each Btu pushed out for cooling in summer must be pushed, on the average, 250 feet. In a fully decentralized system, that traveling distance is more like 2.5 feet—a saving in energy used, for traveling only, of 99%.

3. *Transferring Energy*—For our purpose, Sir Isaac Newton's second law says that when energy (Btu's) are transferred from one body to another, none of them get destroyed, but some of them become unavailable. In short, some always get lost in the shuffle. And the more times they are transferred, the more get lost. In a central system, while heating, the Btu's in the fuel go to the flame, then to the boiler water, then to the air of the room

being heated—three transfers. While it is cooling, they reverse the trip. They go from the warm room air to chilled water in a coil; then to refrigerant in the machine; then to condenser water and, finally, via the cooling tower to the outdoor air—four transfers, for a total of *seven*.

In a decentralized system, while heating, the Btu's go directly from the electric resistance element into the room air—one transfer. On cooling, they go from the room air to the refrigerant, then directly to the outdoor air—two transfers, for a total of *three*. This eliminates four out of seven transfers and results in a saving in transfer losses of 57%. For semi-central systems, the total number of transfers is five, for a saving in transfer losses of 28%.

4. *Partial Load Operation*—Typically, central systems are designed to provide comfort during weather extremes. Thus, most of them operate at full load less than 5% of the time, at partial load over 95% of the time and at less than half load around 70% of the time. Yet, it is well known that boilers and compressors lose efficiency at partial load. And to operate the auxiliaries of a typical system—pumps, fans, cooling tower— may take 0.375 hp per ton of cooling capacity at full load. If so, they take twice this *per ton* or 0.75 hp at half load; 1.5 hp per ton at quarter load and so on. Why? Because these auxiliaries always operate at full capacity. In short, at partial load, the energy utilization efficiency of a central system declines.

In contrast, by cycling on and off under thermostatic control, the major elements of a decentralized system always operate at full capacity. This virtually eliminates partial load losses.

There is a fifth factor that adds to the already high operating costs of many central systems. It applies to all those in the four-season category except the four-pipe systems. I refer to *energy-bucking* design. What does that mean?

When you pay money to simultaneously heat one stream of air or water and cool another, then mix them, you waste energy. One extreme example is a glorious 65° day in June. In the office it would be ideal—75°—with the air-conditioning system off. But this is a dual duct system and it's running. The warm duct, we will say, is delivering 100 cfm of 90° air to the mixing box serving that office. The cold duct, to get the 75° temperature for which the thermostat is set, is delivering 100 cfm of 60° air. When the two streams are mixed, their temperature is 90 plus 60 divided by two or 75°.

Thus, the air-conditioning system has done nothing useful, except provide ventilating air. Yet, heating and cooling are being paid for, even though one cancels out the other completely. That is energy bucking. In more technical terms, we say that the energy utilization efficiency of a fuel-fired central system is low. The larger the building, the lower it is. For heating, let us say it's 40%. This is high for a large building. That means, for every Btu of heat required in the occupied spaces, 2.5 Btu's must be bought and paid for. These come from oil or gas to create the heat and from electricity with which to pump it around.

The energy utilization efficiency of the all-electric decentralized system is the highest so far achieved in practice. For heating a large building, it is usually above 99%. As already stated, anything in the comfort field that can be done with gas or oil can also be done with electricity. But if you use electricity as you *must* use gas or oil in a large building, by burning it under one central boiler, do what the gas company suggests—forget it! You will be compounding a felony if you tie the limitations and inefficiency of a central system to the high cost of electricity.

Yet it has been done—in one monumental new office building in Chicago, for example.

CAVEAT EMPTOR

The components of central systems are designed and produced by engineering-oriented manufacturers who specialize in heavy duty equipment with a 20 to 25-year life expectancy. Window units are now in the hands of household appliance makers or those who, to stay in the game, have been forced to emulate their practices. Their object is high volume production of a standard product, designed for lowest cost and relatively early replacement. The *packaged terminal air-conditioner* or, as it is often called, the *incremental conditioner,* is the major component of the decentralized comfort system for multiroom buildings. It stands at the interface between these two contradictory business philosophies. Both types of firms make it.

Is it any wonder, then, that its spectrum of basic quality— noise level, life expectancy and other vital features—is broad? Actually, it is broader than that of any other basic component of multiroom comfort systems. Thus, its selection calls for the utmost discrimination if you, as the buyer, wish to avoid the penalties and stigma of a *cheap job.* It is the PTAC that will convert the kilowatts you buy into personal comfort and satisfied occupants. If the equipment is noisy, the system is noisy. If it is unreliable, the system is unreliable. And if it has a short life expectancy, so does the major portion of your air-conditioning investment.

Short life expectancy is an inborn weakness of PTAC's built to household appliance standards. Why? Because they should be, but are not, built with the extra stamina needed to withstand

the corrosive effects of rain, snow, smoke and smog, to which—
unlike the household refrigerator—the most costly part of the
mechanism is constantly exposed.

How, then, do you separate the sheep from the goats? How do
you distinguish between the conversational superiority that all
American products possess versus the demonstrable superiority
that is becoming increasingly rare? Price is one but not the only
indication. Comparative net weights also indicate something.
Visiting jobs that have been operating at least a year is highly
useful. So is a visit to the factory where the equipment is being
made. You lose nothing by giving it a careful personal inspection. Asking questions is always useful. What to ask? The following comparisons give the extremes, between which there are all
gradations of quality. They may give you some suggestions for
questions to ask:

TYPICAL APPLIANCE TYPE	HIGHEST QUALITY PTAC

1. Noise Level

Usually in the range of NC-55 on HIGH. Somewhat less on low.	Small sizes satisfy NC-35, large ones NC-40, on cooling. Less on heating.

2. Ventilation

Depends on the vagaries of the wind and building stack effect. Dampers are hand-operated, usually leak in winter. Often left open when unit is stopped, running up heating bills.	Prefiltered, preconditioned air is introduced under positive pressure. Influence of wind velocity and stack effect is minimal. Motor-operated damper closes tight when equipment is off.

3. Condensate Disposal

Condensate produced dur-	Air from centrifugal blower

ing cooling is sprayed by a slinger ring on propeller type condenser fan. Some hits the hot condenser and is evaporated. The rest drips down the face of the building.

atomizes all condensate and delivers it over entire face of condenser. This significantly reduces power input. Equipment is guaranteed not to drip condensate.

4. Life Expectancy

Limited by thin sheet metal that rusts through and heavily loaded motors that fail prematurely—in six to ten years.

Stainless steel fastenings; 16 and 14-gauge, zinc-coated, plastic-enrobed steel; aluminum blower wheels; lightly loaded motors give 20 to 25-year life expectancy.

5. The Icing Problem

Unit may form ice on cooling coil at outdoor temperatures below 70° and still satisfy AHAM standards. Ice blocks air, stops cooling, tends to overheat the compressor motor and may cause its failure.

A unique refrigerant flow control permits equipment to operate at outdoor temperatures as low as 35° without icing. As the outside temperature drops, cooling capacity increases but power input drops.

6. Maintenance

Entire heating and cooling mechanism is normally in one piece, must all be removed in case of failure or for preventive maintenance. This encourages an attempt to repair the equipment right on the job. Sometimes this can be done, sometimes not. In any case, a

All functional components can be quickly removed by non-technical personnel for repair or preventive maintenance. This work is usually done by qualified men at a properly equipped authorized warranty station. Owners normally carry 1% spares for hospital room or executive's office is not the proper place to do it.

service replacement. The repaired parts then become spares.

7. Controls

Controls are standard. There is no choice. They usually consist of a thermostat knob, damper knob and push-button switch for COOL, HEAT and OFF. Provision for high and low fan speed is usually included.

Besides the standard controls, several alternative and extra controls and functions are available: automatic changeover control; provision for automation; central guest room control; continuous vs. intermittant fan operation on heating; reheat (page 215), etc.

8. Warranties

These are usually for one year on the entire unit plus four more years on either the hermetic compressor only or the entire hermetic circuit— compressor, evaporator, condenser, connecting tubing. They may or may not include repair labor. Read carefully to make sure what is covered.

Covers the entire hermetic circuit, including repair labor at the warranty station and recharging with refrigerant. At least one manufacturer guarantees spare parts availability for 20 years from date of purchase. And some include one year's free service in the purchase price.

The configurations in which PTAC's of high quality are produced are likewise quite different than those of the familiar, box-like room air-conditioner design. They are offered in arrangements to suit various kinds of building construction. Generally they are wider and *thinner,* that is, they have a considerably smaller front to back dimension than does a typical room air-conditioner. While they may be largely concealed in the wall, they do not extend beyond its outdoor face. Increasingly, architects have succeeded in treating the facades of buildings, in which they are installed, in a way that enhances rather than detracts from their outdoor appearance.

There are other advantages to the purchase of equipment of the highest quality. It is not an off-the-shelf item. Room cabinets and the enclosures built into the wall to hold the components are made to suit the building in which they are to be installed. Some of the finest installations use attractively-designed, continuous enclosures along exterior walls, on a modular basis. Book cases and storage cabinets may then occupy the spaces between the air-conditioning mechanisms.

But it will take an effort to find and get equipment of the highest quality. Many businesses are today divisions of huge conglomerates. These divisions are often operated by executives who have no experience in and do not understand the business they are operating. Neither does their top corporate management. Thus they have no choice but to operate *by the numbers.* Not understanding the business, they cannot see its long term potential. Required to make a *showing,* they sacrifice the future for instant profit. They do not know how to wear the mantle of leadership in the industry of which they are a part. Thus, wittingly or unwittingly, they feel more comfortable letting their competitors make their decisions for them. Result? Designs tend to become standardized, each maker strives to produce his product at a lower cost than do his competitors, quality suffers.

A large corporation that had a demonstrably superior PTAC for some years switched a couple of years ago to a much cheaper *just as good* imitation of household appliance design. The excellent reputation of the previous design, combined with the low price of the new one, gave it a temporary sales advantage.

That advantage has since been dissipated. Two other leaders in the same field are now engaged in design programs which appear to be headed in the same direction. The classic expression *let the buyer beware* is not obsolete.

DIVERSIFICATION FACTOR

You are debating between a central system and one that is decentralized, for a medium or large sized building. Someone says, "Yeah, but with a central system you can cut the total tonnage." Is it true? It certainly is.

With a decentralized system, you need enough tonnage on the east face of the building to counteract the morning sun. You need enough on the west face to counteract the afternoon sun. And you need enough on the south to offset the low hanging sun on a mild winter day. All together, for the entire building, let us say this amounts to a total cooling capacity of 500 tons.

The central system man says he can do it with a 400-ton machine—despite the losses he has in his distributing system. How come? Because, in the morning, when the east face requires cooling, the west face doesn't—or requires very little. The converse holds true in the afternoon. And, in winter, only the south face will need it. So the central system capacity is delivered to where it is needed and the required total is less than the sum of all maximum loads because they are never simultaneous.

Does this save operating cost? Not at all because only what is needed at any one time is delivered by either system—and central system pumps and fans, for the entire building, are operating constantly, even though local controls may not be calling for either heating or cooling. Total connected horsepower gives a better comparison than total available tonnage. This means not just the compressor but *all* auxiliaries, including the cooling tower. The decentralized system usually shows less connected horsepower than any central system except, possibly, two-pipe fan-coil.

CONFERENCE ROOMS

A problem area, frequently encountered in office buildings, is conference rooms. Whether on the perimeter or in the core, the comfort system was probably laid out for one person per 100 square feet of floor space with ventilation to match. So, it is decided to make a conference room out of what would otherwise have been a private office. Instead of one person per 100 sq. ft. it has, during meetings, three or four times that number—and they all smoke.

Don't blame the comfort system if the air is foul. Triple the air quantity, if it can be done. If the room is on the perimeter and the ventilating air from the perimeter system can not be increased, supplement it by means of a branch duct from the core system and try for 30 cfm per person, total. But if they are bankers, with big black cigars, make it 40 cfm.

REHEAT FOR HUMIDITY CONTROL

In Chapter 11, I described the terminal reheat system. It uses heat to achieve the desired room temperature by having it counteract excessive cooling. But there is another type of reheat. It is to overcome complaint No. 3, that cold, clammy feeling, mentioned near the end of Chapter No. 1. Remember our Florida apartment? The thermometer read 75°. But we felt sticky and clammy all over. The relative humidity must have been over 80%. We needed dehumidification. So we turned on our air-conditioning (decentralized heat recovery system). It gave us what we wanted—dehumidification. It also gave us what we did *not want*—cooling.

What we needed was reheat—enough of it to cancel out the cooling we didn't want. Then our system would only have been dehumidifiying. Thus, the reheat would not be for temperature control but only for humidity control. Where there is a high internal humidity load from people, as in theaters, central systems have used the equivalent of reheat for many years. It has, at times, also been built into package units and even high priced room coolers. Along the Gulf Coast is where it is needed most— chiefly in the evening. Where it is needed, it is a highly desirable feature.

How to do it? One way is to put an electric resistance element in the conditioned air stream. This is simple and has low first cost. But some people are sensitive about spending money for both cooling and heating, at the same time, if they know about it. Water-cooled equipment can feed the warm water leaving the condenser through a fin-coil in the conditioned air stream. In package equipment, hot compressor discharge gas may be diverted to a reheat coil in the discharge air stream.

If you experience that cold, clammy feeling you will now realize why, and that something can be done about it.

BREATHING WALL CONSTRUCTION

PTAC's are normally air-cooled. Thus, they require an outside air opening. Consulting engineers, particularly if they are central system oriented, have been known to say: "The architect won't want those holes in the wall." This ignores the fact that every window is a hole in the wall—many times larger than needed for a PTAC. It ignores the fact that this is not the engineer's problem—or province. The problem belongs in the hands of the man whose job it is to solve it—the architect. Where this has been done, many really beautiful designs have, as already indicated, been the result. As the renowned architect LeCorbusier has said: "'The entire history of architecture is concerned, exclusively, with wall openings."

It is unlikely that he had dogs and chickens in mind. But it should be noted they get rid of their internal heat at one point: the chicken with an open beak; the panting dog with a dripping tongue. Neither can sweat. Humans have sweat glands. They dissipate excess heat from all surfaces of the body. Why, then, shouldn't buildings be designed like humans instead of dogs? There is a lot to be said in favor of *breathing wall construction.*

The fully decentralized, all-electric system requires this breathing wall construction. The decentralized heat recovery system does not. But both offer the same comfort capabilities and both do it at or near the lowest possible cost. So think of the magic possessed by those who have such a system, if it is of the highest quality. At the push of a button, the snap of a switch or the call of a thermostat, it will provide, constantly and immediately, at a full 100% of installed capacity, each principal occupant's own idea of the utmost in full-time personal comfort.

IT'S ONLY MONEY

Chapter 2 contains some suggestions for getting more comfort per dollar. Here are listed a few more that have a direct bearing on first cost, operating cost, or both:

1. The less on-site labor, the lower the cost and, usually, the better the work.

2. The farther your system has to pump air and/or water, the higher the operating cost per unit of capacity.

3. Pushing air through ducts takes more power than pumping water through pipes, for the same amount of heating or cooling.

4. High velocity air ducts or conduits save space and reduce first cost. But they greatly increase power consumption.

5. The fewer the heat transfers, the lower the operating cost and, usually, the lower the first cost also.

6. The lower the temperature lift, the lower the operating cost on cooling. This favors the use of air-cooled condensers except, possibly, in the very hottest, driest part of our country.

7. The less the partial load operation and the less the standby time, the lower the operating cost.

A few years ago a brilliant executive was in great demand as a speaker for sales and other conventions. One of his talks was wrapped around the acronym KISS. And, in due course, his audience learned that this stood for *keep it simple, stupid!* When it comes to comfort systems, this is good advice to observe. My own suggestion to design engineers is, as mentioned earlier, to *Simplicate, Comfort Add and Cost Reduce.* But, over 300 years ago, one of the first of the world's great scientists, Sir Isaac Newton, said it best: "The ultimate in sophistication is simplicity itself."

Chapter 13

Personal Comfort — 1986

The Census Bureau gives our country's present population as 205 million. Did you know that 72%, or 147,000,000 of us, will still be alive in 1986? That is what they say. And that makes living conditions over the next 15 years important to most of us. We all like comfort, we do not all like air-conditioning. It is a safe bet that, during the next 15 years, none of us will experience Utopia, all of us will have our problems. A more comfortable environment will not solve them all. But it can be a positive factor in mitigating them and making life more enjoyable. It can do this where we live and learn, work and play, rest, relax and recuperate.

Most of us, wherever we are, whatever we do, will spend 80% of our time in an indoor man-made environment. The building that provides that environment is expected to outlive its 30-year mortgage. And the better it satisfies its occupants, the better will it be able to compete with what is new and the longer it will hold its value. Therefore, if you want a better environment, the least you can do is to exert your influence in that direction.

Your influence? Yes, the influence we all possess—even if unused—when and wherever we both earn money and spend it;

whether for haircuts or housing, travel or taxes. When a new structure is built or old one modernized, those that will use it can influence what is done. But, for this influence to evoke long time benefits, it is not enough to consider the conditions of today. We need to look ahead and judge what changes are likely to occur tomorrow. This calls for perspective that can only come from comparing where we are now with where we have come from. So let us compare where we were fifteen years ago with the present. This may, at least, give us a clue to where we may be fifteen years from now in 1986.

It is not my purpose to adopt the precarious pose of a prophet. Instead, I invite you to join with me in making a few comparisons between 1956 and 1971. Then, I will invite your attention to prophecies and forecasts which authorities in various fields have recently made—fields which will have a growing impact on indoor environmental conditions of the future.

WHAT'S BEEN HAPPENING SINCE 1956?

It is a moot point whether we humans are making discernable progress in the field of human relations. But, there can be no doubt that, in the field of science and technology, our country is continuing to make rapid progress. Let us, then, consider just ten items—most of which, but not all, have an influence on our environment:

1. In 1956, air pollution was largely a matter of indifference on the part of both government and the public. Now it is not.

2. In 1956, jet air travel was unknown. Top transport speeds were 300 miles per hour. Now they are 600.

3. American per capita electric consumption in 1956 was 4000 kwh. Now it is over 8000—thanks particularly to air-conditioning.

4. Oil and gas desulphurizing plants were unknown 15 years ago. Now several of each are in operation, more under construction.

5. The word Xerox didn't exist in 1956. That year a little-known company, Haloid Corporation, had sales of less than $24 million. In 1970, the same company's sales were 63 times as large and Xerox was in the dictionary.

6. The total capacity of all refrigerating machines in use for human comfort was 30 million tons in 1956. Now it is 75 million tons.

7. Our population was 167 million in 1956. It is 205 million now.

8. No decentralized all-electric comfort system for multiroom buildings was then on the market or in operation. Now, thousands are in use from coast-to-coast in both the USA and Canada, and elsewhere.

9. No decentralized all-electric heat recovery system was then on the market. (Patents on a fuel-fired version were granted in 1955.) Now it, too, is in use from coast-to-coast.

10. Until 1961, no human being had ever been in space. Since then, eight Americans have been seen walking on the moon. They all returned to earth to tell about it.

It is obvious that progress in the field of environmental control cannot possibly be compared with the dramatic accomplishments made in such disparate fields as photocopies and space travel. But, there has been progress and there not only will, but must, be more.

WILL INDOOR COMFORT IMPROVE?

An increasingly powerful nationwide effort is now underway to combat air pollution. For the next few years, the best we can hope for is a holding action. We should consider it successful if we prevent an increase in air contamination beyond the levels that now exist. By then, hopefully, the trend will be reversed and the quality of the outdoor air we now breathe in urban areas will be improving. In contrast, the present progress in improving our indoor environment will continue without interruption. Several factors will spur this favorable trend.

The other day I had a visit with my attorney in the new office into which he had moved a few weeks ago. He told me a story. Mentioning their names, he told of friends who, getting older, had just done what he had done a year or so ago. They had sold their house, to move into an apartment. After looking at everything available in Syracuse—some lovely places—they were now moving into his building. Why?

They gave the comfort system as the compelling reason. It was all-electric, decentralized. Here they would be in full control of their own environment at all times. No other apartment could give them this advantage. Incidentally, what kind of a comfort system served the office my attorney left? Two-pipe induction. What does his new office have? All-electric, decentralized. These people are not unique.

What is the point of this story? Developers are smart. They understand interbuilding competition. They get the message. Systems that provide full-time comfort will drive out those that do not. It is sort of a Gresham's Law in reverse, from which the public can only benefit.

In 1925, when the hungry passengers rushed into the Needles Harvey House, summer cooling was, to them, a miracle. (See Introduction.) Now it is taken for granted. The constantly increasing sophistication of the American public will force higher comfort standards. Between now and 1986, fewer and fewer comfort systems will give the occupant what someone else thinks he should have; more and more will let him have what he wants.

Even in the comfort field, there is no moratorium on technological progress. Just the other day I learned of a new development, now on the market. It promises, for certain applications, to provide more comfort for less money. Thanks to technological progress and related factors, the cost of providing full-time comfort for the occupants of a large, new multiroom building is less today than it was 15 years ago. Technological progress will continue to be made.

Will the government take a hand? It already does, in publicly financed projects such as hospitals and public housing. And a bill has been introduced in Congress to set comfort standards for apartment houses, but only in the District of Columbia—so far. If that one passes, others will follow, as consumerism runs rampant. If such bills are limited to results—in terms of the comfort functions most people want most—they may accelerate the trend toward improved indoor environment. But, if they try to tell the builder how to do it, the results may not be good. They may be no better than they now are in new buildings that the government has erected for its own use. High costs, limited comfort and perpetuation of the vested interests of the air-conditioning industry are the earmarks of a government job.

But, government action or not, the quality of indoor environmental control is destined to improve. Not only that, the need for it will increase. If we integrate into one conclusion what the experts in fields that relate to this subject say, there can be no other outcome.

CAN MORE ENVIRONMENTAL CONTROL BE EXPECTED?[18]

Of course the answer is *yes,* both indoors and out. Factors to consider, that apply to the indoor environment only, indicate that indoor climate control is still a growth industry. Among the reasons why its growth will, of necessity, be more rapid than that of the economy as a whole are these:

Our population is projected to increase another 54 million, or 26%, in the next 15 years. If the acceptance of year-round climate control were to remain static, its growth would follow that of the population. But the market is far from saturated. This is particularly so in private residences, factories and schools. The inclusion of air-conditioning in new buildings has always put pressure on the owners of old ones to do likewise. Take schools: when a new air-conditioned school is built in one part of town, the pressure to install it in the old school, across the tracks, becomes irresistable.

Not only is our population increasing, it is flocking from the country to urban areas. During the 15 years our population as a whole is increasing 26%, the demographers now predict that the populations of the Baltimore-Washington, Los Angeles, and San Francisco Bay areas will increase at more than three times this rate. Shortage and resulting high cost of land will accelerate the present trend toward multiple dwellings. The relative growth of the detached, single-family dwelling will decline.

High density housing must be of good quality or else it will magnify social problems it was intended to abate. A year-round comfort system will help the architect provide pleasing, convenient, secure, quiet family homes at lower first cost and lower operating costs than can be done in any other way. Outside weather conditions are what make it necessary to heat and cool a home. The greater the area exposed to the outdoors, the more heat and cooling are required.

Take, for example, a one-story house of 1000 square feet. Say it is 25 feet wide by 40 feet long, with an 8-foot ceiling.

Walls and roof exposed to the weather will have an area of 2040 square feet. Put this house in a properly air-conditioned multistory multiple dwelling. Provide the interior rooms with automatically modulated lighting to simulate dawn. (Who wants to get up in the dark?) Expose the 25-foot living room wall to the outdoors—nothing else. Now, what do you have? Only 200 square feet of wall exposed to the outdoors. Heat loss in winter, heat gain in summer are now less than 10% of what they were before.

Outside walls and windows are expensive. So are roofs. Soundproof partitions, ceilings and floors in the multiple dwelling will cost less. Sealed windows, by preventing infiltration, will improve comfort, reduce cleaning, cut heating and cooling costs, improve security and also cost less. Heating-cooling maintenace costs and taxes will, in total, be substantially less than for the detached house—at minimum sacrifice in privacy, if properly designed. As condominiums, such dwellings will be within the financial reach of millions who, otherwise, could not afford their own homes.

But this can only be done with complete environmental control. Night temperatures in the country and suburbs fall 15 to 30 degrees from the highs of a summer day. In the city, this doesn't happen. Massive walls of brick and stone and streets of black asphalt, instead of dew-collecting grass, hold the heat. It's that flywheel effect! That is why, for multiple dwellings of the future, complete year-round climate control will be a necessity.

At the same time, air-conditioning will become increasingly necessary for most other types of buildings that do not have it now. Public demand, and owners with their ear to the ground, will see to that. It is a point too obvious to belabor further.

WHAT ENERGY SOURCE WILL BE USED?

Energy is the raw material of comfort. Thus, it is premature to consider *how* comfort will be provided in 1986 until after we arrive at an energy prognosis for 1986.

But first, what about the present energy shortage? Take electricity: To meet the growing demand our generating capacity has been doubling every ten years. Experts believe it will have to continue to do this until late in the 1970's, after which the rate may decline. This means present capacity must be multiplied by 2.2 to 2.4 times by 1986. But lately, the necessary growth rate has not been met. Completion of new nuclear generating stations has been delayed for two reasons: manufacturing problems and public opposition. Shortage of low pollution fuels has compounded the problem. Nor have high interest rates helped to finance expansion.

The present gas shortage is the direct result of wellhead price fixing by the Federal Power Commission. Prices have been too low to encourage the drilling of enough new wells to meet the demand. Impending price increases should relieve the shortage —temporarily.

Oils used for heating and power generation are the ones in short supply. This applies particularly to residual fuel oil. An unusually heavy demand for this product has been created by shortages of gas and coal, anti-pollution legislation and the delay in starting up new nuclear generating stations.

What may we expect in 1986? Let us see what the experts say:

1. Gas—From 1962 to 1968 our known reserves of natural gas in the USA dropped from a 13.4 to an 11.8 years supply, according to Lawrence O'Connor of the FPC. Since then, reserves have continued to slip. We are using gas faster than we are finding new supplies. A study by the National Academy of Sciences shows U.S. production of natural gas as reaching its all-time high in 1973. After that, it will decline rapidly. By 1986, our annual production will be less than it is today, despite a burgeoning increase in demand.

To make up the deficiency that already exists, three things are being or will be done: (a) Gas imports from Canada will grow, to the extent the Canadians will let us have it. (b) Liquid natural gas (LNG) will be imported by the USA, as it already is by Europe, in special ships, at 259° below zero. Contracts for us to get it from Algeria and Venezuela have already been placed, the ships are now under construction. (c) Synthetic *natural* gas will be made out of coal, by hydrogenating it. Pilot plants for doing this are already in operation.

What will this do to the cost of gas? During the 15 years ending in 1963, its retail price has increased 76.1%. Since then, the increases have continued. Drillers have had to go deeper. The deeper they go, the more it costs per foot. Pipeline gas is now delivered to New York City area gas companies for about 40¢ per million Btu. The landed price of LNG is now estimated at 60 to 65¢ or 50 to 60% more. What about synthetic pipeline gas? We do not know. But El Paso Natural Gas Company, for one, wouldn't have made a 25-year contract with the government of Algeria for LNG if they thought the synthetic product could be made for less in this country.

Some of the experts say the cost of natural gas will go so high that most of it will eventually be used for the production of chemicals. In any case, the price can only go one way—up.

2. *Oil*—Our known reserves of oil are somewhat better than those of gas. The rate at which new reserves are being found, thanks to the North Slope of Alaska, is, so far, keeping pace

with increased production. Known U.S. reserves, at our present rate of usage, should carry us to 1990. But peak U.S. production is again forecast for 1973—possibly 1975. Then it will drop. Increasingly, we will have to depend on other sources.

Oil is easier to import than gas. Much of what we now use comes from abroad. But foreign governments are getting more restrictive. They want a constantly increasing cut of the pie.

The Organization of Petroleum Exporting Countries—six in the Persian Gulf area, plus Libya and Venezuela—are in the driver's seat and they know it. The new five-year agreement, recently concluded between OPEC and the international oil companies, had a predictable result: an immediate large increase in payments to OPEC plus further increases to take effect on January 1 of each year.

Peru recently expropriated the local subsidiary of an American oil company—a practice started by Mexico in 1938. Creeping expropriation puts our oil imports increasingly at the mercy of foreign—not always friendly—governments.

Our retail price of light fuel oil increased 36.9% from 1948 to 1963. It has continued to increase. During the past year, the price of heavy, low-sulphur fuel oil has risen over 50%. Diminishing reserves, deeper wells, longer hauls, off-shore drilling and cost of desulphurizing will all assure higher future prices.

There is now increasing talk in this country of making liquid fuel from coal. It is not new. Germany was forced to do it during the War. And even though liquid fuel now costs two or three times more in coal-rich parts of Europe, as against what it costs here, the practice of making it from coal was dropped there. It costs too much.

3. Coal—The U.S. Geological Survey puts our recoverable

coal reserves at 830 billion tons—a 1500-year supply at present rates of consumption. Another authority says our peak rate of production will not be reached until about the year 2200. Those of us who expect to still be around in 1986, won't have to worry about a coal shortage, if the miners stay on the job.

It is not clear if these figures include the low grade form of coal called *lignite*. High water and ash content, and a heat value half that of good bituminous, makes its long range transportation too costly. Mine-mouth use is indicated. Coal will continue to be our most reliable, economical and abundant source of fossil fuel energy.

4. Nuclear Energy—The energy available in one pound of fissionable uranium is equal to that from 1300 *tons* of coal. Our reserves of uranium and thorium are equivalent to about 160 times our assured reserves of coal, gas and oil combined. Referring to our potential supply of nuclear energy, an authority says, ". . . the uranium and thorium ore in the U.S. . . . must be on the order of hundreds to thousands of times greater than the world's initial supply of fossil fuels."

5. Electricity—The increase in our per capita consumption of electricity from 4000 to 8000 kwh per year during the past 15 years is modest, as against the projected value of 18,500 kwh per person in 1986. Where will it come from? It now is generated from coal, gas, oil and, increasingly, nuclear fuel. Hydroelectric power is now and will continue to be a minor source. As reserves of gas and oil continue to decline, while prices advance, their use for power generation will decline. Oil, and probably LNG, will be increasingly diverted from stationary to mobile and other uses for which they are uniquely suited.

Half of our present coal output is now used to generate electricity. By 1986, over 75% will, it is forecast, be used for this purpose, particularly at mine-mouth and especially in the case of lignite. At present, about 23% of all our raw fuel is converted into electricity. By 1986, this is forecast to be 33%. By then, 14% of our *total* energy usage will be nuclear. Since most of it will be used to generate electricity, over 40% of our electrical energy will come from that source.

What will this electrical energy cost? Electricity is one of the few commodities that has dropped in price during the 15 years ending with 1963—14.7%. Since then it has continued to drop. In 1964, the FPC announced a further reduction of 27% in the cost of electricity as its goal for 1980. Recently, however, the downward trend has reversed itself. Inflation, and exhorbitant interest on money borrowed to expand capacity, have seen to that. Nevertheless, cheaper electric power has, for many years, been an objective of our government. And I haven't forgotten that when I was a boy a kwh of electricity cost more than two loaves of bread. Now it costs less than one slice.

One fact, often overlooked, should be remembered: Every time we use a ton of coal, a cubic foot of gas, a gallon of oil we are, in effect, using up a nonreplaceable source of energy. It took the sun billions of years to put it on earth. It is a finite supply. When it is gone, it will be gone for good.

As the use of nuclear power becomes general, high cost power areas will no longer exist. Why? Because the cost of transporting nuclear fuel is so low, in relation to its generating capacity, that the cost of nuclear power will be essentially the same everywhere.

From the foregoing facts and forecasts we can, now, make a sound prognosis as to the energy that will be used in 1986. While our supplies of gas and oil will decrease, prices will increase.

While our supplies of electricity increase, the price will decrease, if not in actual dollars, at least in terms of a fixed dollar. Hence, this unequivocal prognosis: by 1986, *electricity will be the generally accepted form of energy for both heating and cooling* in this country.

HOW WILL IT BE DONE?

Having made this prognosis, we are now in position to answer the question: "How will it be done?" Or, more accurately, how will the electricity be used for human comfort? As has been amply proven on many large installations, and previously explained and reported here, the operating cost of decentralized all-electric systems is now less than that of central systems that employ combustible fuel as the source of heat. It is much lower than for systems that use electricity at a central point for heating with hot water. Today, more and more new, detached, single-family dwellings have electric heat. In some areas, their operating cost is less than for gas heat. Mostly, however, it is more. But owners consider the extra convenience, cleanliness, reliability and safety worth the extra cost. By 1986, no other source of heat will be as economical as electricity for all structures, *provided the design of the system capitalizes reasonably on what can be done with electricity,* that cannot be done with combustible fuels. This means increased decentralization as the size of the building increases.

Still another factor will encourage this trend. I refer to our water supply. Some municipal jurisdictions have already curtailed the permissable summer operating hours of large air-conditioning systems that use water for condenser cooling. Systems with air-cooled condensers have not been so curtailed. The more a system is decentralized, the more feasible—and economical—it is to use air-cooled condensers.

For these reasons, I am confident that, by 1986, new central and district systems, as we know them today, will no longer be the accepted approach to environmental control in large structures. Instead, irrespective of the total capacities required, the heating and cooling effect will be produced in increments, the capacity of which matches the requirements of the individual spaces being served. While a multiplicity of systems may then serve one large space, owners will no longer buy one large system to serve a multiplicity of spaces.

In short, the dogmatic imperative, that the heating and cooling of a large building should be by means of equipment and apparatus centralized at one spot in the structure, will have gone the way of the steam locomotive and propeller-driven air transport.

WHAT ABOUT TECHNOLOGICAL PROGRESS?

Technological progress will, of course, continue to be made. One area has to do with the fact that electricity cannot be stored, but what it does can be stored. One-fourth of the cost of electric power is for generating it, the rest is for delivering it and standing ready to deliver it to the point of use. Our electric utilities have an average load factor of 64%. That means that they deliver only 64% of the power they could deliver, if running steadily at full capacity. At very little additional cost, they could deliver over 50% more power, thus permitting significant price reductions.

Control systems are now on the market that automatically improve the load factor for many types of large buildings and other applications. These increase load factor, thus reduce operating cost by reducing *demand charge*. Demand charge is what the utility charges for being ready to deliver, at all times, at the maximum rate their customers may require.

A further improvement in the load factor of comfort installations is possible by the use of storage systems. Heating and cooling effect may be stored until needed, during periods of low electric demand: one in the form of hot water, the other as ice. But it may be more economically done—as to power and space —by means of chemical mixtures called eutectic solutions. These can be tailored to suit specific temperature levels. Research on materials of this kind is continuing. Pressure to reduce operating costs and shortage of power calls for higher generating station load factors. Combined, these influences may increase the interest in storage systems.

No real breakthrough during the next 15 years is seriously anticipated in the comfort field. But there is one that has long been hoped for. It has to do with thermo-electric heating and cooling. About 150 years ago Peltier, a Frenchman, discovered that if a direct current is passed through a closed loop of wire, both heating and cooling can be produced. To do this, the loop requires two junctions, each made of dissimilar metals or alloys. When power is turned on, one junction gets hot, the other cold. Reversing the flow of current reverses the temperature of the junctions. The system is compact and has no moving parts. It has fantastic potentialities for convenient all-electric year-round comfort. But don't wait for it.

Thermo-electric systems are in limited use on some of our submarines and in a few other highly special applications. Unfortunately, their energy utilization efficiency, using presently known materials for junctions, is very low. Operating costs can-

not come near to competing with present systems. The continuing
search of science is for a presently unknown thermo-electric
material that will make this a commercially feasible system.

IS AIR POLLUTION A FACTOR?

In our country, space heating contributes only about 5.5%
of the total pollutants in the air—a mere 8.8 million tons a year.
Of this, 3.3 million tons is sulphur dioxide and 2.2 million tons
is carbon monoxide. The rest—over one million tons each—is
evenly divided between nitrous oxides, hydrocarbons and particulates—the stuff you feel when you place your hand on the
sill of an open window.

If you take one thousand average houses, heated with soft
coal, they will put out about 3300 pounds of pollutants per day.
Change to oil and this will drop to 640 pounds. Gas will cut it
to 285 pounds. With electric heat it is nil. So methods of heating
do have a direct bearing on air pollution.

But, you say, all this does is to move the source of pollution
from the individual houses to the electric generating plant. Yes
and no. Neither hydroelectric nor nuclear plants put pollutants
into the air. Fuel-fired plants do. In their favor, however, are
the economies of scale. They get much more energy and less
pollutants out of a pound of coal than any homeowner can.
Then, too, they are increasingly restricted to low-sulphur fuels.
They are easy to locate and control. That is not true with thousands of small users. And, increasingly, the new generating stations are near the mines, away from urban centers. Thus, electric
heat contributes much less to air pollution, in total, than would
fuel-fired houses requiring an equivalent amount of heat.

The swing to all-electric comfort systems is one reason why

an increase of over 200% in the per capita power usage is projected for the next 15 years. The necessary larger generating plants bring up another side of the pollution problem. Emotional environmentalists and the sensation-seeking news media have increasingly fallen in love with a new expression, *thermal pollution*. They have tardily discovered that electric generating stations reject waste heat to rivers, lakes or the ocean. It's been done for years. All fuel-fired plants do it. So do atomic plants. But only the latter are judged guilty without a trial.

To generate a kwh of electricity—equal to 3412 Btu—almost twice that amount of waste heat must be thrown off by the generating station—whether fossil or atomic fuel is used. Taking our present per capita usage of 8000 kwh per year, requires the disposal of enough heat to melt over 175 tons of ice per capita.

Put this heat into water and you do not taint it. You only make it warmer. Such warmed water has been returned to its source by electric generating stations for over 50 years. The thermal effects should, by now, be known. It is reported that seaside power plants cause shrimp and oysters to thrive. I hear that some of the best fishing on Lake Ontario is around the warm discharged water from power plants. It's a nice kind of pollution, since it improves swimming, too.

But, in any case, if you want electricity you must permit the power plant to get rid of its waste heat. Some new plants have been forced to switch from the use of river water, as intended, to water recooling towers. This results in a considerable cost increase and can only increase the cost of power. What happens when this is done?

If they all used cooling towers, what would it do to the atmosphere? In 1970, it would have added 50,000 pounds of humidity (25 tons of water) per capita; in 1986 it would add over 115,000 pounds per capita. Increased humidity causes increased discomfort. Increased humidity requires increased cooling capacity. Increased cooling capacity takes increased power. Larger generating plants breed more humidity: oscillating technological redundancy—with a vengeance.

Where in the world is there a greater source of *thermal pollution* than the Gulf Stream? Yet, what would the British Isles and Scandinavia be like without it? Those who want to talk about thermal effect should think the matter through, before they use the pejorative term *thermal pollution*.

Engineers rarely decide *what* to do, only *how* to do it. Engineers are not opposed to wise technology. But they are opposed to engaging in unwise technology—to satisfy those who lack information on the subject they are talking about. Fortunately, the steps now being taken to improve our *outdoor* environment, especially by reducing air pollution, cannot help but have a favorable impact on our *indoor* environment.

We should not forget that, for countless ages, man was unable to create dwellings that protected him fully from bitter cold, devitalizing heat and stifling humidity. Then, it was considered noble to ignore discomfort, not let it distract one from the task at hand. Or, as Confucius put it reproachfully, 2500 years ago: "The superior man thinks always of virtue; the common man thinks of comfort."

Now, life is more complex, living space is shrinking. We are, today, increasingly subject to stresses that were unknown to those who lived in the placid pastoral days of the past. Thus, more than ever before, each of us owes it to himself to avoid those stresses that are, in fact, avoidable. One such is the stress of physical discomfort. For the first time in man's history, he now has the power to eliminate, while indoors, physical stress entirely.

All that remains is for each of us to resist the forces of tradition, and, using the knowledge we have gained, to insist on indoor environments conducive to stress-free work and play, rest and relaxation. And the more we insist on higher standards of indoor comfort, the sooner will we get them.

Abbreviations

AH&RN	Air Conditioning, Heating & Refrigeration News
AHAM	Association of Home Appliance Manufacturers 20 N. Wacker Drive, Chicago, Ill. 60606
AP	Associated Press
ARI	Air-Conditioning & Refrigeration Institute 1815 N. Fort Myer Drive, Arlington, Va. 22209
ASHRAE	American Society of Heating, Refrigerating & Air-Conditioning Engineers
Btu	British thermal unit(s)
Btuh	Btu per hour
cfm	Cubic feet per minute
cps	Cycles per second
db	Decibel
DD	Degree Days
EAC	Electrostatic Air Cleaner
ET	Effective Temperature
hp	Horsepower
F	Fahrenheit
FPC	Federal Power Commission
Kcal	Kilogram calorie (3.96 Btu)
kw	Kilowatt of electricity
kwh	Kilowatt hour
MBtu	Thousand Btus
NEC	National Electric Code
OA	Outside air
OPEC	Organization of Petroleum Exporting Countries
psi	Pounds per square inch
PTAC	Packaged terminal air conditioner
rh	Relative humidity (percent)
R-11	Refrigerant 11 (Trichloromonofluoromethane)
R-12	Refrigerant 12 (Dichlorodifluoromethane)
R-22	Refrigerant 22 (Chlorodifluoromethane)
SCR	Silicon controlled rectifier

238

Credits and References

It is utterly impossible for me to recall and give credit to all the many sources from which material was used in writing this book. After all, most of it has been accumulating in my head over a span of 47 years. To the extent that I have knowingly used material from others, I wish to give them full credit and express my deep appreciation to these sources.

Such items, unless otherwise indicated, are identified in the text by superscribed numerals, as follows:

1. *American Aviation,* May, 1968. Used by permission.
2. *Appliance Magazine,* May/August, 1969, Dana Chase, Editorial Director. Used by permission.
3. *Air Conditioning, Heating & Refrigeration News,* Frank Versagi, Editor. Used by permission.
4. *Heating/Piping/Air Conditioning,* July, 1963. Used by permission.
5. *Architectural and Engineering News,* April, 1965.
6. *ASHRAE Guide and Data Book, Applications.* Copyright, 1971, by ASHRAE. Used by permission.
7. Material quoted from, based on or influenced by *ASHRAE Handbook of Fundamentals.* Copyright, 1967, by ASHRAE. Used by permission.
8. *THE INCH,* Summer, 1966. Texas Eastern Transmission Corporation. Used by permission.

9. Based on *ASHRAE Guide and Data Book, Equipment.* Copyright by ASHRAE. Used by permission.

10. Margaret Ingels, *Father of Air Conditioning.* Copyright, Carrier Corporation, 1952. Reproduced by permission.

11. Joseph B. Olivieri, P.E., *How to Design Heating-Cooling Comfort Systems.* Copyright Business News Publishing Company. Used by permission.

12. *ASHRAE Guide and Data Book, Systems.* Copyright, 1970, by ASHRAE. Used by permission.

13. Temperatures from *The Weather Handbook,* H. McKinley Conway, Jr., Editor. Copyright 1963. Used by permission of Conway Publications, Inc., Atlanta, Georgia.

14. From *Modern Air Conditioning, Heating and Ventilating,* by Willis H. Carrier, Realto E. Cherne, Walter A. Grant, and William H. Roberts. Copyright, 1959, by them. Copyright, 1940, by Carrier Corporation. Reprinted by permission of Pitman Publishing Corporation.

15. Historical material is, in part, drawn from "The History of Refrigeration; 220 years of Mechanical and Chemical Cold: 1748-1968", Willis R. Woolrich, *ASHRAE Journal,* July, 1969, and "Milestones in Air Conditioning", by Walter A. Grant, *ASHRAE Journal,* September, 1969.

16. Roger W. Babson, *Looking Ahead 50 Years.* Used by permission of Harper & Row, Publishers, Incorporated.

17. J. P. McEvoy, "Pepe Was Never So Peppy!" *Reader's Digest,* October, 1953. Used by permission.

18. Influenced and aided by, "The Human Habitat, 1994", a speech given at the ASHRAE Convention, San Francisco, January 19, 1970, by past president William L. McGrath.

Index

241

244

DATE DUE

APR 1 8 '94			